Current Trends in Magnesium (Mg) Research
http://dx.doi.org/10.5772/intechopen.97999
Edited by Sailaja S. Sunkari

Contributors
Mariana Correa Rossi, Liliana Romero Resendiz, Vicente Amigó Borrás, Fatma Tugce Senberber Dumanli, Fabrizio D'Errico, Martin Tauber, Michael Just, Palanivel Mathiazhagan, S. Jayabharathy, Yolanda Castro, Emilia Merino, Alicia Durán, Sailaja S. Sunkari, Abhineet Verma

First published in London, United Kingdom, 2022 by IntechOpen
IntechOpen is the global imprint of INTECHOPEN LIMITED, registered in England and Wales, registration number: 11086078, 5 Princes Gate Court, London, SW7 2QJ, United Kingdom

British Library Cataloguing-in-Publication Data
A catalogue record for this book is available from the British Library

Additional hard and PDF copies can be obtained from orders@intechopen.com

Current Trends in Magnesium (Mg) Research
Edited by Sailaja S. Sunkari
p. cm.
Print ISBN 978-1-80355-480-8
Online ISBN 978-1-80355-481-5
eBook (PDF) ISBN 978-1-80355-482-2

Meet the editor

Dr. Sailaja S. Sunkari obtained her Ph.D. from the University of Hyderabad, India, in 2003. She was a JSPS postdoctoral fellow (2003–2005) and a JSPS Bridge Fellow (2016) at the Department of Chemistry, University of Tokyo, Japan, and a DST-India Women Scientist (2007–2008) at Banaras Hindu University, India. In 2008, she joined Banaras Hindu University as an assistant professor in the Department of Chemistry, Women's College, where she teaches inorganic chemistry to undergraduate students and is pursuing research in the field of hybrid functional materials. Her research interests include the structure-dependent physical properties (luminescence and magnetism) of supramolecular assemblies.

Contents

Preface

Magnesium (Mg) is an exciting element of the s-block of the modern periodic table, playing a crucial role in photosynthesis. While this is an undisputed aspect of Mg, this element also plays multiple roles as a material in diverse physical forms in a wide variety of industries. This book highlights the diverse aspects of Mg and presents research on its uses with a focus on Mg-based products for sustainable living.

The book is divided into two sections. Section 1 includes two chapters that outline the applicability of Mg-based products in diverse sectors, with due consideration to environmental impact. Section 2 includes four chapters that discuss fine-tuning the parameters of Mg-based alloys/composites/compounds that are critical for their use as engineering or biofunctionalized materials.

Chapter 1, "Introductory chapter: Magnesium - A Perspective" by Abhineet Verma and Sailaja S. Sunkari summarizes some of the important sectors where Mg is involved, including transportation, energy storage, and the chemical industry.

Chapter 2, "Magnesium Alloys for Sustainable Weight-Saving Approach: A Brief Market Overview, New Trends and Perspectives" by Fabrizio D'Errico, Martin Tauber, and Michael Just discuss the environmental impact of Mg processing technologies and the high costs involved in bringing the material into general use. Discussing the factors that drive the demand for Mg alloys in the transportation industry, the authors address the concerns of environmental impact due to processing technologies and provide a hopeful perspective on how the market can propel with Mg-based products for sustainable living.

Chapter 3, "Magnesium in Synthesis of Porous and Biofunctionalized Metallic Materials" by Mariana Correa Rossi, Liliana Romero Resendiz, and Vicente Amigó Borrás presents the advantages of the inherent characteristics of Mg that make it a suitable candidate to couple with other elements for alloying. The chapter deliberates on the experimental techniques to combine Mg with titanium (Ti) for achieving biocompatible Ti–Mg alloys suited for cell growth required for bone degeneration or for use in bone implants.

Chapter 4, "The Role of Silane Sol-Gel Coatings on the Corrosion Protection of Magnesium Alloys" by Emilia Merino, Alicia Durán, and Yolanda Castro examines the relevance of silica sol-gel coatings as an alternative method to control the corrosion of Mg and its alloys and discusses strategies for modification of the sol-gel matrix by incorporation of different types of inhibitors to achieve an active barrier property in Mg alloys for corrosion prevention. Mg alloys are widely explored as structural components of materials in aerospace, automobile, and biomedical fields due to their lightweight and biocompatibility. However, for any material to display its efficient function, its stability is paramount under experimental working conditions. The need for such stability becomes more vital when the functional environment is biological media.

Chapter 5, "Experimental Investigation of Mechanical and Wear Behaviour of AZ91 Magnesium Hybrid Composite Materials" by P. Mathiazhagan and S. Jayabharathy critically examines the significance of hybrid Mg composites and details the development of new and cost-effective magnesium-based alloy hybrid materials by addition of reinforcements like titanium, titanium dioxide, and graphene via stir casting technique for use in varied industrial applications. In the present consumer-driven world, hybrid composite materials overweigh normal materials in terms of fulfilling consumer requirements with minimal harmful effects on the environment.

Chapter 6, "Magnesium Borates: The Relationship between the Characteristics, Properties, and Novel Technologies" by Fatma Tugce Senberberr Dumanli illustrates the emerging synthetic methodologies to obtain magnesium borates, with redesigned morphologies finding use in adsorption, ion-battery agents, and hydrogen release agents in chemotherapy applications. The chapter highlights the benefits of novel methods like solid-state or hybrid synthetic methods over traditional methods in obtaining Mg borates, a starting reagent of extensive use in several industrial applications, and highlights the relationship between their synthesis, characteristics, and novel applications.

Present-day Mg research opens new frontiers in materials research addressing the major concerns of sustainable living and gives hope of retaining a healthy planet earth for future generations.

I wish to thank the chapter authors for their excellent contributions. I am also grateful to Author Service Manager Ms. Marina Dusevic at IntechOpen for her guidance and patient help throughout the publication process. I offer sincere thanks to IntechOpen for the invitation to be the academic editor of this book. I also thank Prof. Satyen Saha and Mr. Abhineet Verma for their help. Finally, I thank my parent institution Banaras Hindu University, India, and the IoE incentive grant for senior faculty that helped me take part in this book project.

Sailaja S. Sunkari
Department of Chemistry,
Mahila Maha Vidyalaya,
Banaras Hindu University,
Varanasi, India

Section 1

Magnesium - A Multifaceted Metal

Chapter 1

Introductory Chapter: Magnesium - A Perspective

Abhineet Verma and Sailaja S. Sunkari

1. Introduction

Magnesium (Mg) with atomic number 12 is situated in the II group and III period of the modern periodic table. Discovered in 1755 by *Joseph Black*, Mg derives its name from Magnesia, a district in Eastern Thessaly in Greece [1]. Introduced to a middle school student in the form of Mg wires burning with bright white flame (luminescent), Mg is a solid at RT (**Figure 1**), soft metal with a m.p. of 650°C and a b.p. of 1090°C and is the eighth-most abundant element in earth's crust.

Mg is well studied for its excellent properties in diverse fields as in land/air transportation, energy storage devices, catalysis, medical implants, food, nutrition, etc. (**Figure 2**).

2. In transportation

One of the characteristic features of Mg, that makes it most explored in diverse fields, is its low density. With a density of 1.74 g/cc, it is lighter than aluminum, Al (2.7 g/cc). The low density of Mg and its alloys, rendered it to be studied extensively in the transportation industry, viz., automobiles and aerospace.

Traditionally, developments in magnesium alloys have been driven by industries like automobiles and aerospace requirements, for any material which is lightweight to operate with the huge demands of the industries (as in engines, gearbox, or casings of aircraft). The low density, only two-thirds that of aluminum, and dramatic improvements in anti-corrosion performance and mechanical properties in recent years have been attractive to designers to use Mg alloys in aerospace applications [2, 3]. Such alloys/components are now specified on projects like the McDonnell Douglas MD 500 helicopter/B52 Stratofortess (**Figure 3**). There are many applications of alloys of magnesium in aerospace industries [4, 5], gearbox parts in a helicopter (Westland Sea King) and wheels of aircraft, both in ZW3. Magnesium when it is forged, is also used in aircraft engine applications. In the future, magnesium forgings are most likely to be used in higher temperature applications.

Besides the aerospace industry, another industry that has enjoyed the advantage of the low density of Mg is the automotive industry. In the 1920s, magnesium began to appear in the automotive industry. The lightweight metal began to be used in racing cars to add to their competitive edge in MRI153M alloy (Mg-Al-Ca-Sr based alloy).

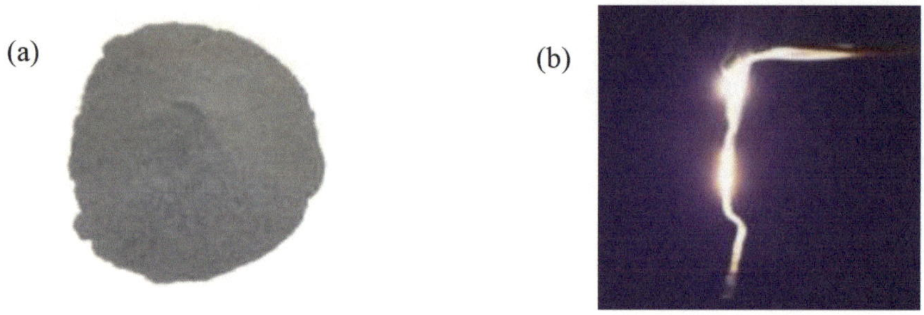

Figure 1.
(a) Magnesium powder (b) Burning magnesium wire.

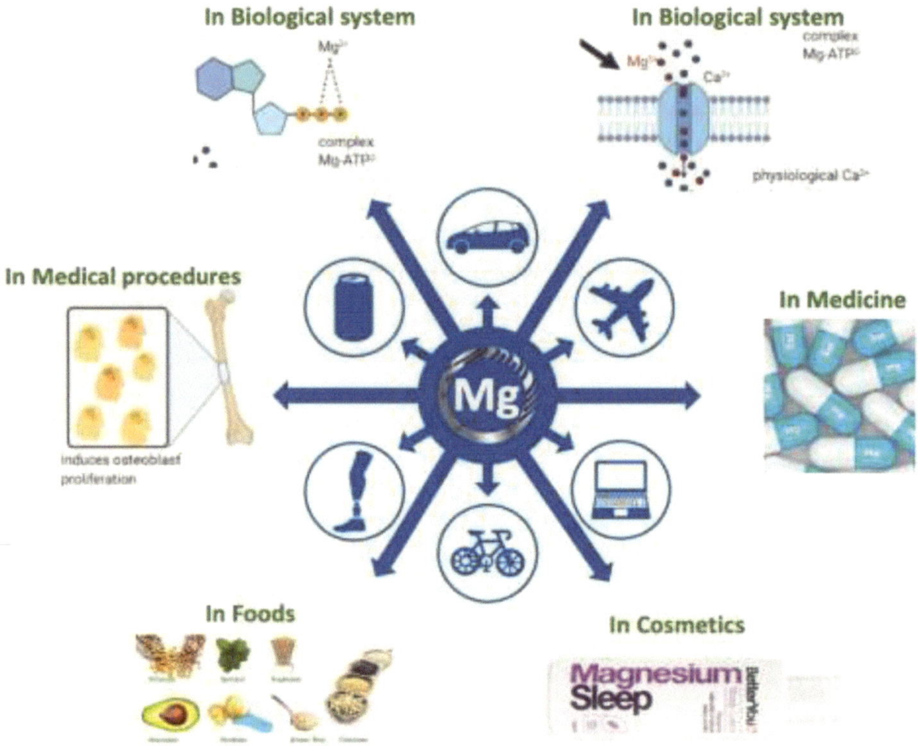

Figure 2.
Schematic diagram showing varied applications of magnesium (Mg).

Years later, commercial vehicles such as Volkswagen Beetle started using magnesium-containing about 20kg (44.09 lbs) of total weight. The interest in magnesium use in automotive applications has increased over the past ten years due to the increasing environmental and legislative influences [7, 8]. Magnesium contributing to increased fuel efficiency, accelerated performance, and sustainability has led to their use in vehicles. Many automotive companies have found magnesium to be a suitable replacement for steel and aluminum, for their products (**Figure 4**). Audi, Daimler (Mercedes-Benz), Ford, Jaguar, Fiat, and Kia Motors Corporation are just a few. Magnesium is currently used in gearbox, front end and IP beams, steering column, driver's airbag housings, steering wheels, seat frames, and fuel tank covers (AJ62A (98.8–91.8% Mg) and Magnox (Al 80))·[6].

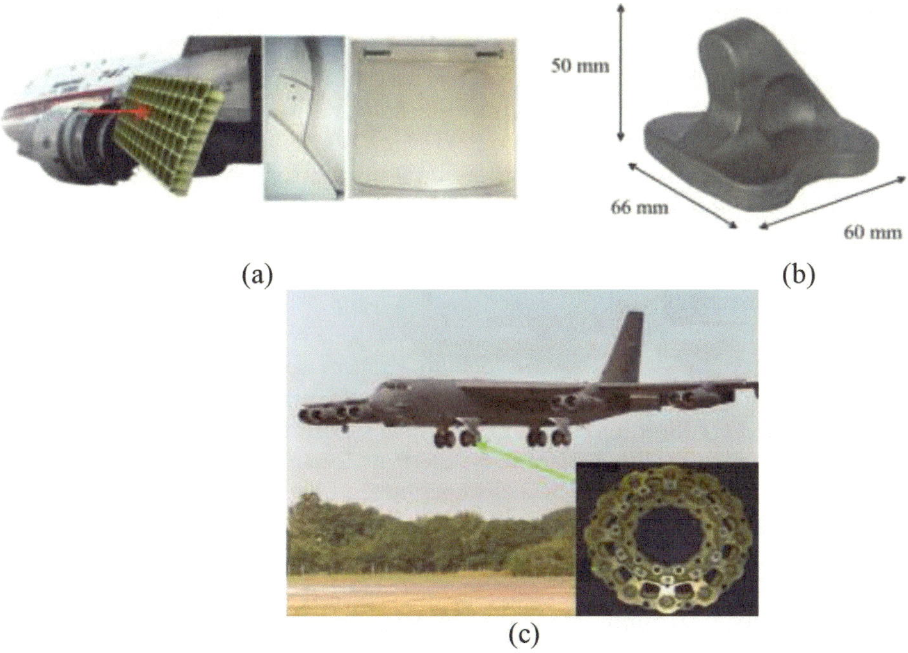

Figure 3.
*Examples of application for Mg alloys in (a) Boeing 747, wings, and seat components. (b) Aircraft door lock set made of magnesium alloy [7] (**c**) B-52s fighter-jet (wheel bearing) [6].*

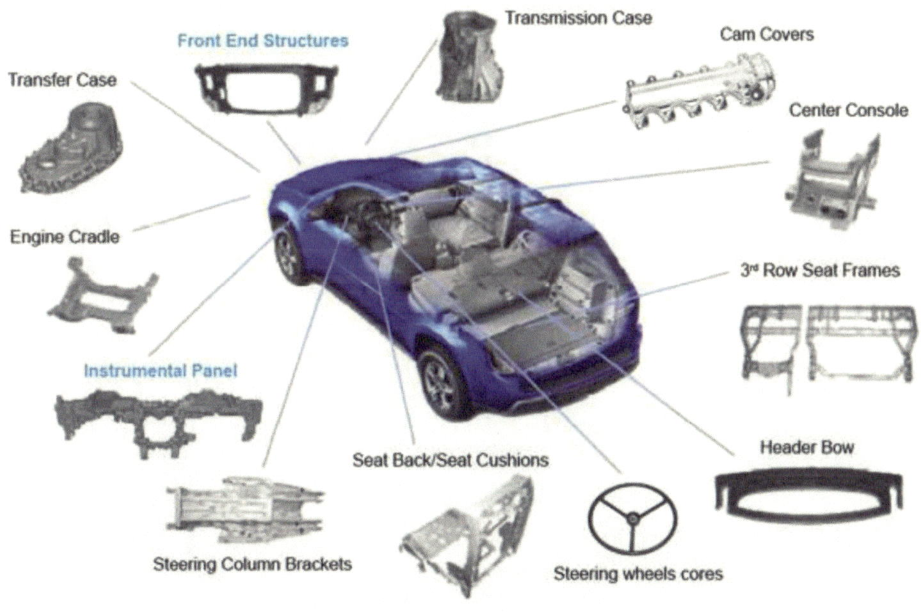

Figure 4.
Examples of applications for magnesium alloys in the automotive industry [8].

Using magnesium in automotive applications helps more than just weight savings. For a number of years, the desire to identify challenges, search for solutions, and explore new opportunities regarding the use of magnesium in vehicles has been growing. Magnesium usage on the front end provides a lower overall mass for the car. It allows for shifting the center of gravity towards the rear of the car, improving handling and turning capabilities. In addition, frequencies that reduce vibration and overall noise can be achieved by tuning magnesium-containing parts. A single cast piece of magnesium is preferably used to replace steel components in vehicles, thus providing material with additional strength and allowing housings to be cast into place. This castability involves the merits of low manufacturing cost and less tooling and gauges.

3. In medical industry

Not only lighter but Mg also form excellent alloys with Al and other metals due to its excellent mechanical, welding, and fabrication capabilities. Moreover, such alloys are widely studied for their improved mechanical strength, corrosion withstanding capabilities, etc., thus enabling them to be used as biocompatible alloys, energy storage devices, etc.

The Use of Mg-based products in the form of composites, and alloys in the medical industry is far widely studied. Ease of corrosion inhibition in the physiological conditions enabled exploration of Magnesium (Mg)-based biocomposites and alloys to be used in biomedical applications such as bone fixation, cardiovascular stents, hip joints, screws/pins, dental implants, etc. (**Figure 5**) In the Mg-based composites, the matrix materials are biomedical magnesium alloys based on Mg–Ca, Mg-Al, Mg–Zn, and Mg–REE alloy and the reinforcements are based on hydroxyapatite (HAP), calcium polyphosphate (CPP), and β-tricalcium phosphate (β-TCP) particles [9].

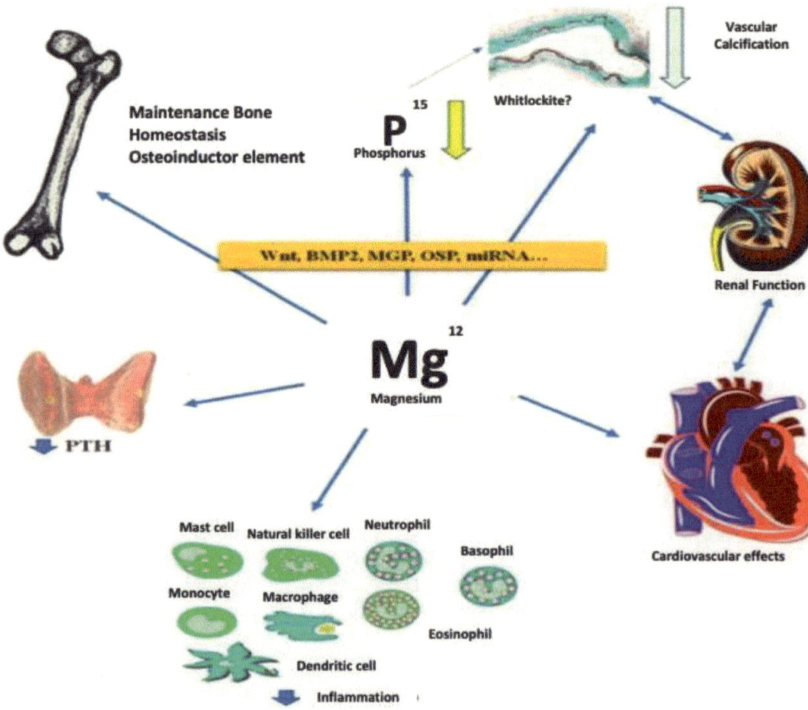

Figure 5.
Application of Mg-based products in Medical Implants.

4. In catalysis

Magnesium oxide is one of the most important metal oxides in catalysis. Though commonly used as support, its employment as a catalyst was also reported in *viz.*, oxidative coupling of methane [10, 11], dehydrogenation–dehydration of alcohols [12], dehydrohalogenation of halogenated hydrocarbons [13], benzylation of aromatics [14], synthesis of pyranopyrazole derivatives [15], and Claisen–Schmidt condensation, etc. [16] Furthermore, the catalytic performance of MgO was improved by promotion by other metals [17] and addition of small amounts of iodine to the reactant mixture [14] for dehydrogenation of short-chain alkanes.

Using alkoxy magnesium $Mg(OR)_2$ and dialkyl magnesium $Mg(alkyl)_2$ species as Grignard reagents in organic synthesis is a fundamental textbook example of organic synthesis [12, 18]. However, the utilization of magnesium as a catalyst in asymmetric synthesis is dramatically undeveloped compared to that of almost all other transitional metals, despite its relatively higher natural abundance than all these other metals combined. Magnesium catalysts are widely used in chemical transformations, and their catalytic use in asymmetric synthesis is highly advisable. For enantioselective reactions, magnesium salts such as $Mg(OTf)_2$, $Mg(NTf_2)_2$, and $Mg(ClO_4)_2$ are used as Lewis acid catalysts. Different magnesium reagents can give rise to chiral magnesium catalysts. One method involves coordinating complex formation using chiral ligands and a strong Lewis acid magnesium salt, described as "the magnesium catalytic strategy with a fixed magnesium salt" [19].

5. In energy storage

Magnesium is under research in the energy sector as a possible replacement or improvement on lithium-ion batteries in certain applications. In the early 1990s, the potential use of magnesium batteries was recognized based on V_2O_5, TiS_2, or Ti_2S_4 cathode materials and magnesium metal anodes [20]. Magnesium has a (theoretical) energy density per unit mass under half that of lithium (18.8 MJ/kg vs 42.3 MJ/kg) in comparison to that of lithium as an anode material. Still, a volumetric energy density is around 50% higher (32.731 GJ/m^3 vs 22.569 GJ/m^3), making its use advantageous [21]. Magnesium anodes do not exhibit dendrite formation compared to metallic lithium, except in certain nonaqueous solutions and at current densities below ca. 1 mA/cm^2. Such dendrite-free Mg deposition allows magnesium metal to be used without an intercalation compound at the anode, thus raising the theoretical maximum relative volumetric energy density to around five times that of a lithium graphite electrode [22]. Additionally, modeling and cell analysis have indicated that magnesium-based batteries may have a cost advantage over lithium due to the abundant magnesium on earth and the scarcity of lithium deposits [20, 23].

Another viable emerging energy storage technology under research is the Magnesium-Sulfur rechargeable battery (**Figure 6**) that uses magnesium ion as its charge carrier, magnesium metal as the anode, and sulfur paste as a cathode. To increase the electronic conductivity of the cathode, sulfur is usually mixed with carbon to form a cathode composite. Currently, efforts on rechargeable magnesium battery research are underway at Apple, Toyota, and Pellion Technologies [6] and in several universities.

Successful research outcome would be of great interest since, in theory, the Mg/S chemistry can provide 1722 Wh/kg energy density with a voltage of ~1.7 V, addressing most future energy challenges.

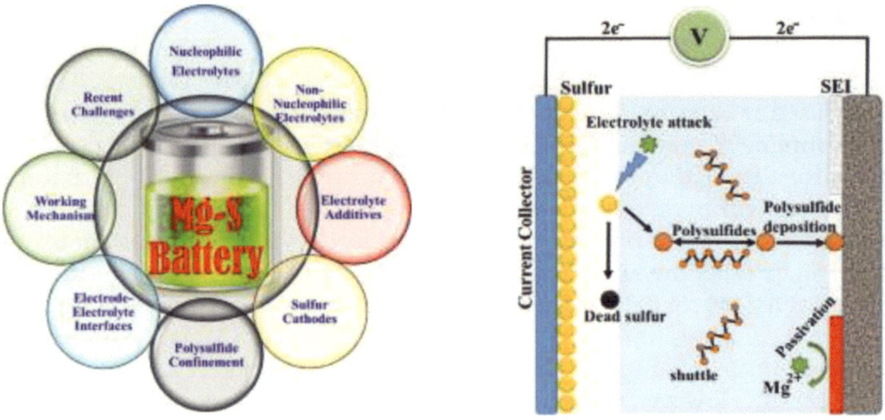

Figure 6.
Schematic diagram of Mg-S batteries showing the working principle and formation of magnesium-polysulfides (Mg-PS), which passivate the anode surface [20].

6. In thermoluminescence

Use of magnesium silicates and borates (for example, MgO, Cr^{3+}/ Be-doped MgO, $Mg(SiO_3)_n$, $MgO(B_2O_3)_3$, Li_2O-MgO-B_2O_3, $Sr_3MgSi_2O_8$, etc.) as thermoluminescent materials are reported [24]. In this phenomenon exhibited by certain crystalline materials, previously absorbed energy from electromagnetic radiation or other ionizing radiation is re-emitted as light upon heating the material [25]. TL glow curves for 2 mg $Sr_3MgSi_2O_8$ substance exposed to radiographic radiation by 254 nm UV source show a resolute single peak around 123°C [26]. A glow curve deconvolution technique was used for analyzing acquired glow curves. Moreover, magnesium orthosilicate intensity of TL glows curves shows a dose-response linear relationship extending to 20 Gy. To a certain degree, TL features of this phosphor are conditional beside Mg_2SiO_4 is suggested to consider a suitable substance for dosimetry [27].

7. In food and nutrition

Magnesium is the fourth most abundant metal ion, showing important physiological functions in the human body. Magnesium balance is maintained by renal regulation of magnesium reabsorption. According to the United States Food and Nutrition Board recommended daily allowance for magnesium is 420 mg for adult males and 320 mg for adult females, respectively. Green vegetables, nuts, seeds, and unprocessed cereals are rich sources of magnesium; besides, some magnesium is made available in fruits, fish, meat, and milk products. However, consumption of processed foods, demineralized water, and agricultural practices using soil deficient in magnesium for growing food, lowers the consumption levels than the recommended amounts, resulting in magnesium deficiency leading to hypomagnesemia, cardiac or neurological disorders [28].

8. Conclusions

Magnesium displays its utility in diverse fields ranging from materials to medical fields, as discussed in the preceding sections. Improvements in processing technologies have wide scope for increased use of materials based on magnesium, which may revolutionize the transportation sector in terms of energy and cost savings. Corrosion withstanding capabilities in biological media, improved mechanical strength, and low density make Mg and its alloys to be used as potential candidates in medical implants. Not only in medical and material fields, but the contribution of Mg in the energy sector is also no less, as Mg batteries are being explored as an alternative to Li ion batteries. Similarly, improved catalytic efficiencies based on Mg catalysts are advantageous for chemical industries too. In the coming years, sustainable life is not far away with the involvement of magnesium and its products in almost all fields of daily life.

Acknowledgements

AV like to thank CSIR SRF (09/013(0825)/2018-EMR-I) for financial support. SSS gratefully acknowledges IoE incentive for Sr. Faculty (Dev. Scheme No. 6031), Banaras Hindu University, Varanasi, India.

Author details

Abhineet Verma and Sailaja S. Sunkari*
Department of Chemistry, Institute of Science, Banaras Hindu University, India

*Address all correspondence to: sunkari.s7@gmail.com

References

[1] Segal D. Materials for the 21st century. Oxford: Oxford University Press; 2017

[2] Zhang J, Kim J, Zargaran A, Hwang J, Suh B, Kim N. Designing a magnesium alloy with high strength and high formability. Nature Communications. 2018;**9**(1):1-6

[3] Kurzynowski T, Pawlak A, Smolina I. The potential of SLM technology for processing magnesium alloys in aerospace industry. Archives of Civil and Mechanical Engineering. 2020;**20**:23

[4] Kulekci MK. Magnesium and its alloys applications in automotive industry. International Journal of Advanced Manufacturing Technology. 2008;**39**:851-865

[5] Pekguleryuz M, Kainer K, Kaya A. Fundamentals of Magnesium Alloy Metallurgy. Now Delhi: Woodhead Publishing; 2013

[6] Dziubińska A, Gontarz A, Dziubiński M, Barszcz M. The forming of magnesium alloy forgings for aircraft and automotive applications. Advanced Science Technology Research Journal. 2016;**10**(31):158-168

[7] Henn Y, Fein A. Project MagForming: Development of new magnesium forming technologies for the aeronautics industry. Publishable Final Activity Report. 2010:1-58

[8] Gwynne B, Lyon B. Magnesium alloys in aerospace applications, Past Concerns, Current Solutions Magnesium. In: Triennial International Aircraft Fire & Cabin Safety Research Conference. 2007

[9] Bommala V, Krishna M, Rao C. Magnesium matrix composites for biomedical applications: A review. Journal of Manganese Alloy. 2019;**7**(1):72-79

[10] Hargreaves SJ, Hutchings GJ, Yoyner RW, Kiely CJ. The relationship between catalyst morphology and performance in the oxidative coupling of methane. Journal of Catalysis. 1992;**135**(2):576-595

[11] Choudhary V, Rane V, Gadre R. Influence of precursors used in preparation of Mgo on its surface properties and catalytic activity in oxidative coupling of methane. Journal of Catalysis. 1994;**145**(2):300-311

[12] Aramendıia M, Borau V, Jiménez C, Marinas J, Porras A, Urbano F. Magnesium oxides as basic catalysts for organic processes. Journal of Catalysis. 1996;**161**(2):829-838

[13] Mishakov I, Bedilo A, Richards R, Chesnokov V, Volodin A, Zaikovskii V, et al. Nanocrystalline MgO as a dehydrohalogenation catalyst. Journal of Catalysis. 2002;**206**(1):40-48

[14] Choudary B, Mulukutla R, Klabunde K. Benzylation of aromatic compounds with different crystallites of MgO. Journal of the American Chemical Society. 2003;**125**(8):2020-2021

[15] Babaie M, Sheibani H. Nanosized magnesium oxide as a highly effective heterogeneous base catalyst for the rapid synthesis of pyranopyrazoles via a tandem four-component reaction. Arabian Journal of Chemistry. 2011;**4**(2):159-162

[16] Patil A, Bhanage B. Novel and green approach for the nanocrystalline magnesium oxide synthesis and its

catalytic performance in Claisen–Schmidt condensation. Catalysis Communications. 2013;**36**:79-83

[17] Morales E, Lunsford JH. Oxidative dehydrogenation of ethane over a lithium-promoted magnesium oxide catalyst. Journal of Catalysis. 1989;**118**(1):255-265

[18] Morrison R, Boyd R, Bhattacharjee S. Organic Chemsitry. New Delhi: Pearson; 2011

[19] Dunski H, Jowiak W, Sugier H. Dehydroxylation of the surface of magnesium oxide by temperature programmed desorption. Journal of Catalysis. 1994;**146**(1):166-172

[20] Rashad M, Asif M, Ali Z. Quest for magnesium-sulfur batteries: Current challenges in electrolytes and cathode materials developments. Coordination Chemistry Reviews. 2020;**415**:213312

[21] Aurbach D, Gizbar H, Schechter A, Chusid O, Gottlieb H, Gofer Y, et al. Electrolyte solutions for rechargeable magnesium batteries based on organomagnesium chloroaluminate complexes. Journal of the Electrochemical Society. 2001;**149**(2):115-121

[22] Liebenow C, Yang Z, Lobitz P. The electrodeposition of magnesium using solutions of organomagnesium halides, amidomagnesium halides and magnesium organoborates. Electrochemistry Communications. 2000;**2**(9):641-645

[23] Kim H, Arthur T, Allred G, Zajicek J, Newman J, Rodnyansky A, et al. Structure and compatibility of a magnesium electrolyte with a sulphur cathode. Nature Communications. 2011;**2**(1)

[24] Abass M, Diab H, Abou-Mesalam M. New improved thermoluminescence magnesium silicate material for clinical dosimetry. Silicon. 2021;**14**(6):2555-2563

[25] Azorin J. Preparation methods of thermoluminescent materials for dosimetric applications: An overview. Applied Radiation and Isotopes. 2014;**83**:187-191

[26] Dewangan P, Bisen D, Brahme N, Sharma S, Tamrakar R, Sahu I. Thermoluminescence glow curve for UV induced Sr3mgsi2o8 phosphor with its structural characterization. Journal of Materials Science: Materials in Electronics. 2018;**30**(1):771-777

[27] Dogan T, Akça S, Yüksel M, Kucuk N, Ayvacikli M, Karabulut Y, et al. Comparative studies on thermoluminescence characteristics of non-doped Mg_2SiO_4 pre- pared via a solid-state reaction technique and wet-chemical method: An unusual heating rate dependence. Journal of Alloys and Compounds. 2019;**795**:261-268

[28] Swaminathan R. Magnesium metabolism and its disorders. Clinical Biochemist Reviews. 2003;**24**(2):47-66

Chapter 2

Magnesium Alloys for Sustainable Weight-Saving Approach: A Brief Market Overview, New Trends, and Perspectives

Fabrizio D'Errico, Martin Tauber and Michael Just

Abstract

In the transportation sector, weight-saving strategies emphasize greenhouse gas reductions by improving fuel efficiency. Furthermore, it is a fact that consumers appreciate less-consuming vehicles. Lighter battery electric vehicles (BEV) mean higher travel distance covered with the same battery charge. Still, the fuel range of BEV is today not a secondary issue for choosing an e-vehicle as a unique family vehicle. Weight-saving strategies are also a priority for hydrogen gas-fuelled vehicles. Until hydrogen fuel for the transport sector is not produced at affordable costs in fully renewable pathways, increased fuel efficiency is critical for the product appeal. Magnesium is an environmentally compatible and biodegradable material with a similar density to structural plastics. On the contrary, plastics are responsible for nonbiodegradable microplastics in deep-marine environments when not recycled or correctly treated at their end of life. Due to the costly usage of lightweight materials, priority is given to activities to reduce costs by developing new materials and increasing the affordability of manufacturing costs. In this chapter, magnesium is presented from much perspective point of view: we will base it on comprehension of the past, considering the present, but with some ambition to propel hearts over today's obstacles.

Keywords: magnesium, automotive, weight-saving, carbon footprint, lightweight, electrolytic, Pidgeon

1. Introduction

The need for weight-saving in the automotive and mass transportation sector, like trains and civil airplanes, has historically pushed the usage of magnesium, which, for shared knowledge, is the metallic material at the lowest density, nearly to dense polymers. But, much more effectively, magnesium alloys are characterized by very high specific strength. A long tradition and past knowledge of the magnesium industry accumulated from the 1970s till its Golden Age in the early

1980s. In those years, you could buy primary magnesium at its lowest price on the marketplace, and many bet that the turn against rival aluminum was just around the corner. However, the forecast high growth rate of the magnesium market has not succeeded yet.

Today, you can hear about big worries about magnesium:

- It could be an unsafe material, susceptible to easy burning and explosion.
- It has a high cost and poor availability on the market.
- It has a high carbon footprint in the extractive, refining, and casting processes.

You could also add to the list a poor knowledge of the wrought alloys and their deformation processes and concerns about their poor corrosion and creep resistance. Those barriers prevented magnesium from competing with its main rival in weight-saving strategies in the transport sector, the aluminum metal. In the following, we try to give readers a more detailed view, considering that we have to know what we were in the past to get a keen comprehension of today's concerns.

Most concerns about the magnesium market do not depend on geographical lack of raw material. Still, trade issues, production base, and export policies made primary production in Europe not competitive. The last primary production plant in Europe shut down in 2001 since European-based smelters could not compete with low-cost Chinese production. As a result, the availability of primary material is a genuine concern, as European demand depends mainly on China's imports. Therefore, one main drawback for broader use in the automotive industry is the lack of a solid supply base with stable prices over a medium-term period combined with competitive magnesium production outside China. The last primary production plant in Europe shut down in 2001 since European-based smelters could not compete with low-cost Chinese production. As a result, the availability of primary material is a genuine concern, as European demand depends mainly on China's imports. Therefore, one main drawback for broader use in the automotive industry is the lack of a solid supply base with stable prices over a medium-term period combined with competitive magnesium production outside China. But few people know which milestones were in the history of the magnesium market. Why did magnesium growth not meet reasonable expectations? Why did the material price increase and fluctuate after prolonged stability at the lowest price level targeted in the 1980s, the years of maximum Western production? What shaped today's market structure based on perilous dependency on Chinese producers? In the following, we'll try to give you a compass to never get lost in such a multifaced and complex market.

2. Magnesium demand in the transportation industry: learning from the past about historical barriers to magnesium growing demand

During First World War, Americans noticed the importance of magnesium for its strategic pyrotechnics application. Magnesium was the base of flares incorporated in rocket devices that, fired into the air, descended with a parachute, lighting the

DOI: http://dx.doi.org/10.5772/intechopen.102777

enemy's corridor for several minutes. During the Interwar period, 1919–1939, the interest in strategic magnesium for national armaments industries rose worldwide (**Table 1**). The rise in the magnesium demand was pushed by lightweight structural applications. Alliances were surprised by the German Luftwaffe supremacy of the burning European skies. German airplanes were faster and capable of carrying unexpected bomb shipments. By studying some German planes that crashed, the British discovered that they contained a large percentage of magnesium alloys, the "Elektron metal" as the Germans called it. The weight-saving in German aircraft was the key to such a significant advantage in the European skies. Magnesium was instantly proclaimed as a strategic metal for the second time. The U.S. Government allocated

Starting year	Process	Sources	Company	Region	Type
1895	I.G. Farbenindustrie	Seawater/brine	I.G. Farbenindustrie	Germany	Electrolytic
1915	Dow process	Seawater/brine	Dow Chemical	Midland, Michigan (USA)	Electrolytic
1920	Dow process	Fluoride material and magnesium oxide	American Magnesium Corporation (Alcoa)	California (USA)	Electrolytic
1920	Dow process	Seawater/brine	De Norske Saltverker AS	Bergen, Norway	Electrolytic
1931	I.G. Farbenindustrie's technology	Dolomite/FeSi	National AluminiumMagnesium Institute (VAMI) USSR	Leningrad	
1933	I.G. Farbenindustrie's technology	Dolomite/FeSi	Riken Metal Manufacturing Co.	Ube (Japan)	Thermal
1935	I.G. Farbenindustrie's technology	Dolomite/FeSi	Government plant	Zaporozhye and Solikamsk (USSR)	Electrolytic
1936	I.G. Farbenindustrie's technology	Dolomite/FeSi	Magnesium Electron Company (MEL)	United Kingdom	Electrolytic
1941	Dow process	Seawater	Dow	Freeport, Texas (USA)	Electrolytic
1945	I.G. Farbenindustrie's technology	Brucite Mg (OH)2	Aluminum Company of Canada (Alcan)	Arvida (Quebec)	Electrolytic
1951	I.G. Farbenindustrie's technology	Seawater	Norsk Hydro	Porsgrunn, Norway	Electrolytic
1959	Pidgeon	Dolomite/FeSi	Alabama Metallurgical Corp.	Selma, Alabama (USA)	Thermal

Starting year	Process	Sources	Company	Region	Type
1960	Pidgeon	Dolomite/FeSi	Furukawa Magnesium Corp.	Oyama (Japan)	Thermal
1964	Magnetherm	Dolomite/FeSi	Pechiney	Marignac (France)	Thermal
1964	Pidgeon	Dolomite/FeSi	Ube Kosan	Ube (Japan)	Thermal
1969	Modified IG I.G. Farbenindustrie's technology	Brine	National Lead Industries (From 1980, facility operated by Amax Inc.; from 1989, facility operated by Magnesium Corp. of America. MagCorp)	Great Salt Lake, Utah (USA)	Electrolytic
1970	Modified IG I.G. Farbenindustrie's technology	Dolomite/FeSi	AM Magnesium	Texas (USA)	Electrolytic
1972	Dow process	seawater	Dow Chemical	Texas (USA)	Electrolytic
1974	Amati-Ravelli	Dolomite/FeSi	Magnesium do Brasil	Ceara (Brasil)	Thermal
1975	Magnetherm	Dolomite/FeSi	Alcoa's Northwest Alloys	Washington (USA)	Thermal
1992	Norsk Hydro	Magnesite	Norsk Hydro	Quebec (Canada)	Electrolytic
1993	VAMI/ UTI Technology (Russian)	Brine	Dead Sea magnesium	Israel	Electrolytic
1994	Alcan	Asbestos	Noranda	Canada	Electrolytic
1997	Alcan	Magnesite	Queensland Metals Corporation Limited	Australia	Electrolytic

Table 1.
History of worldwide magnesium plants before 2000s.

all of the U.S. nation's total production (at that time produced by Dow Chemical) to national defense. At the beginning of the Second World War, the production of magnesium was 33,500 tons, whereas 5 years later, magnesium production reached a peak of 426,000 tons [1].

Americans developed their own wrought and cast magnesium alloys. Enormous quantities of magnesium were put on military aircraft to curb the weight of liquid and air-cooled engines, wheels, oil tanks, frame structures, instrument housings, gyro frames, and many others. The jet-propelled prototype "Flying Wing" airplane was an aircraft bomber, designed for high speed and maneuverability, made primarily of magnesium (**Figure 1a**). It never entered service in favor of the B-36 bomber (**Figure 1b**) that used a total of 3800 kg of magnesium in castings, forgings, and sheets for airframe parts ad fuselage skin. At the same time, for civilian scope, commercial truck vehicle, body, and motor engine parts, benefited as well from the light-weighting that was made possible by magnesium. Magnesium alloys were extensively used in the airframe skin of the large airplane Convair XC-99 built by the U.S. Air force that remained in activity from the 1940s to 1950s. By 1948, the military aircraft Lockheed F-80C

DOI: http://dx.doi.org/10.5772/intechopen.102777

(a) (b)

Figure 1.
(a) The "Flying Wing" airplane and (b) the B-36 bomber airplane.

Figure 2.
The Lockheed F-80C is constructed of magnesium throughout, today visible at the United States Air Force Museum, Ohio.

"Shooting Star" was the first American project for constructing a combat-ready jet fighter capable of exceeding 500 mph in level flight. One F-80C (47-171) constructed magnesium throughout, redesignated NF-80C-LO, is today visible at the United States Air Force Museum, Ohio (**Figure 2**).

However, following the end of the Second World War, military applications of magnesium lost their strategic importance. The magnesium extractive industry contracted to register a new peak demand in the early '50s because of the Korean War.

The production peak registered during Second World War drawn by U.S. national production was not surpassed until the '70s. Widespread post-war applications of magnesium would be expected in automobiles and civil aircraft to reduce engine weight and dynamic masses. Still, magnesium demand finally decreased till the '70s, not being sustained by aggressive market growth strategies. Magnesium soon revealed losing in front of the prominence of aluminum alloys. The significant factor restricting the growth of magnesium after wartimes can be researched—as a comparison—looking to a good lesson taught by the rival aluminum industry. The primary aluminum industry had a long tradition of cooperation. A group of pioneers in the European aluminum industry set up an "Aluminum Association" way back in 1901, just 15 years after the modern electrolysis smelting process patents of Charles Martin Hall and Paul Héroult. It was created to promote the widespread use of aluminum (at that time, aluminum was a pioneering material for few applications) to provide economic governance to the nascent aluminum market [2]. The Aluminum Association shared information on markets, feedback from customers on applications, on the quality of the metal. All those information was necessary to align demand requirements and supply features and to encourage the private sector's investments. It was an observatory to analyze the market trends to make the use of aluminum alloys easier. Moreover, the Aluminum Association directed specific actions toward pricing policy based on stable selling prices to promote demand growth. This stability consolidated a nonspeculative market, and it allowed to plan a gradual and programmable extension of the productive capacities of big plants.

These efforts were not completely replicated in the nascent magnesium market to sustain post-war demand. It is true that a prominent American producer, the Dow company, broadened civilian markets by the '1950s. Precisely, the date 1954 was when the Dow company started the mass production of Samsonite Ultralite luggage bag, 20% lighter than other luggage bags, entering in operativity a mammoth large-scale magnesium sheet mill. At that time, several advancements in magnesium alloys were made. New coatings (anodic, electroless-Ni, and Cr-plating) were produced in the 1950s to protect the magnesium alloys from corrosion; viable ceramic and porcelain coatings for magnesium were developed, processes for cladding magnesium sheet and plate alloys with other magnesium alloys and aluminum.

At Dow company, people frequently told that Dow's metallurgists within the 1960s probably had alloyed magnesium with any possible element with good wettability like Li, Al, Si, Ca, Mn, Cu, Zn, Sr, Y, Zr, Ag, and Rare Earth [3]. Researchers soon focused on the overall properties of a fabricated Mg-alloy component as a result of microstructure [4] finally realized by alloy chemistry and processing parameters to promote beneficial solid solution distribution, dispersoids, intermetallic precipitation by heat treatments, grain refining.

Corrosion behavior of Mg alloys developed was enormously improved by limiting impurities Fe, Ni, and Cu that largely influence corrosion resistance of Mg because of the formation (and dispersion) of micro-galvanic cells. New Mg-RE, Mg-Th, and Mg-Th-Zr high-temperature alloys were developed at the beginning of the '60s for use at temperatures of 200–350°C but were limited to their high costs to jet aircraft and military missiles. Following the first hot chamber die-casting process developed at Dow Chemical Company [5], further die-casting techniques were improved and widely used to make engine-driven tools (chain saws, post hole diggers, etc.). Researchers and metallurgy laboratories at magnesium companies provided many answers to questions about phase equilibria, alloying effects, and the relationship of structure and properties for their potential customers (casters, forgers, extruders). During the 1960s in Europe, 20,000–25,000 tons, supplied mainly by Norwegian Norsk Hydro, were being used in the Volkswagen Beetle's air-cooled engine and gearbox. Those components were installed above and behind the rear wheels, and this required the German engineers to produce a drive system as light as possible so that the front wheels gripped the road adequately. The 1960s were also the Cold War years, and several magnesium sheets were used in the lightweight intercontinental ballistic missiles. A machined magnesium-lithium alloy LA 141 was chosen for its high stiffness, low weight, and sound vibration damping characteristics for manufacturing the chassis of the Launch Vehicle Digital Computer (LVDC) that provided the autopilot for the Saturn V, the liquid-fueled rocket developed under the Apollo program for human

Figure 3.
The Gemini spacecraft with the centered, white-painted portion in magnesium alloy.

exploration of the Moon. High-temperature magnesium-thorium alloys in sheet and extrusion form constituted a large part of the large conical structure of the Retro-Rocket Modules of the Gemini spacecraft (the white-painted portions in **Figure 3**, just near the black-painted cone).

However, it was symbolic of what president Roger Wheeler said at the 23rd Annual Meeting of the Magnesium Association (still, International Magnesium Association) in 1966. He said that the magnesium industry had failed in the previous 15 years to take its place as a fundamental industrial commodity metal in the U.S. [6]. At that time, magnesium consumption was one order less than forecast one decade earlier. The Magnesium Association recognized that the future of magnesium growth in North America could depend on the automotive market, and automotive engineers needed to lose their concerns about magnesium, following the example of Germans [7]. In Germany, likewise the air-cooled Volkswagen Beetle die-cast engines, in 1967 Porsche developed the 166 kg crankcase for their six-cylinder 911 series (**Figure 4a**), following visionary forecast in the post-wartime (**Figure 4b**).

By the 1970s, developments were extended to new composite magnesium-based materials, new high-temperature magnesium alloys, new fluxing methods, squeeze casting, recycling methods, and new anodizing processes for magnesium [8]. They were good news for the magnesium industry, but, in reality, in the middle 1970s, world demand for magnesium was about not more than 2% of the aluminum. Economic uncertainties by the oil crisis caused the rapid decrease in sales of the Beetle caused German automakers to curtail magnesium consumption [9]. Despite aluminum pricing that remained steady, the rising magnesium price made aluminum much more competitive. Whether during the late 1970s and beginning 1980s, the need for cutting fuel cost of automobiles could represent an opportunity for the magnesium industry, desulphurization and deoxidation of steel were (unfortunately) still considered the most favorable growth market for magnesium [10]. Magnesium for aluminum alloying was expanding market sector as it consumed almost half of the magnesium production, and it was expected to grow at about 5% per year, while some other markets such as desulphurization and die casting were expected to grow more rapidly [10]. Many efforts were made by researchers to develop high-performance alloys for automotive applications to curb as much weight as possible from massive

(a) (b)

Figure 4.
(a) The Porsche 166 kg crankcase for the six-cylinder 911 series; (b) the November 1944 issue of Light Metal Age presented an image of "Tomorrow's light metal car." In the associated article, the writers talk about the use of aluminum and magnesium in the sleek-lined, spaced-aged cars of the future.

engine blocks, including advanced rapidly solidified magnesium alloys [11] and magnesium alloy composites [12].

Over the second half of the 1980s and early 1990s years, the period was a time of great ferment for magnesium potentialities in the automotive sector [8]. The dominant technology for magnesium production was still electrolysis with giant plants, and the leading producer countries were the United States, Canada, and Norway. Die casting consumption with different magnesium die-cast components in automobile sectors drove the significant annual growth rate of North America—thus more or less the total world magnesium demand at that time. General Motors die-casted in a single-shot, a large part an instrument panel beam for the GMC Savana and Chevrolet Express van. It was a 12 kg part 4 mm thick, which provided 32% mass saving compared to the steel design with improved crashworthiness and high vibration damping. It was less costly thanks to a few parts assemblies, 25 parts in the magnesium design compared to the 67-parts in the steel-made product [13]. To shape much more complex parts made of magnesium, in 1991, the Dow Company registered the Patent for a method and apparatus for the injection molding of magnesium metal, a process based on the foundation of the fundamental discoveries on semi-solid metals by Flemings and his students at Massachusetts Institute of Technology, MIT [14].

During that golden age for magnesium, the US Dow Company increased its almost monopolistic control of the magnesium market thanks to the economy of scale of its giant and old electrolytic plants powered by low-cost power sources available in Texas. In 1991 Dow could produce 109,000 tons per year, namely around 35% of the world's entire output. But the cost of making magnesium in Texas began to rise gradually as the time of cheap natural gas ended. With almost 20 kW-h of electricity to produce a kilogram of magnesium, a lot of power was available, but all that power had been contracted for by the big aluminum producers like Alcoa, Kaiser, and Reynolds [3]. The old Dow plant became soon antiquated, and to stay in business at a competitive level, the most significant World producer would have had to build a new efficient plant, as planned at the Great Salt Lake, a project that never started. Factors leading to Dow's success, and that driven till the early 1990s all magnesium market, have been: early entry, cost efficiency, and strategic deterrence behavior [3]. The biggest world's magnesium producer started to hand over its 60 years of harvests by the early 1970s when Dow began to reap the benefits of its magnesium business rather than investing beyond its old plants in Texas. Dow company switched from a "limit pricing" strategy originally designed to deter entry to a "skim pricing" type of strategy that ultimately sacrificed the firm's viability as a magnesium producer [15].

Unlike Dow, other dominant firms have opted to expand tactically in related industries (e.g., DuPont in titanium dioxide and Alcoa in aluminum before 1945). One potential explanation is that Dow's cost advantage was not sustainable. Dow's production process benefited from years of incremental improvements but was not fundamentally different from the technology potentially available to others [15]. The Dow big electrolytic plants worked at an efficient scale in the decades after wartime characterized by modest demand for magnesium, and there were substantially no further opportunities for new efficient-scale plants until the U.S. But a radical change, as depicted, started with an automotive interest in magnesium at the beginning of the 1990s. Magnesium would switch its position in the marketplace from a specialty material with one dominant producer with considerable knowledge accumulated in 60 years into a commodity product with a competitive global market [15].

The rest of the story is like what happened to dominant Western countries firms in similar markets for commodity products. As the Cold War ended around 1990 and as the Chinese economic reform entered its Second Stage (the establishment of the Socialist Market Economy), individual Chinese enterprises were allowed to exist and to be protected by the law of the People's Republic of China. The primary market forces began to shift very rapidly in Western countries. In China, a multiplicity of low-investment production plants with the Pidgeon process were building at the minor technology scale. Hundreds of those plants based on a very high labor-intensive process were set up and ramped up very rapidly, in a few months, producing per capita just a few hundred metric tons per year. There was no Chinese knowledge at that time about magnesium alloying and alloys applications; those plants needed just to sell primary magnesium to the Western countries at almost their marginal cost. This new situation created confusion in the not-stabilized marketplace [3]. However, it is a fact that the Pidgeon process produced a significant amount of World War II magnesium. Those Pidgeon plants during wartime could not compete with the electrolytic process.

On the contrary, in the 1990s, when small Chinese plants started to supply 4% of the world's magnesium, Chinese labor cost was very low. In that period, Chinese-made magnesium was sold at about 0.72 USD/lb. while Dow's production cost was not less than 1.08 USD/lb. That magnesium price had been starting to crush the marketplace, a problem that never ended till that time. By far, the principal use of magnesium, almost 50%, that year was recorded in alloying the large numbers in the aluminum beverage cans sector, in which magnesium was (and is) used as a strengthening agent instead of in structural alloys for engineering applications.

Very soon, China, with its low prices, supplied 50% of World magnesium demand (**Figure 5**), becoming the world's largest supplier of primary magnesium. Between 2000 and 2010, magnesium production in China tripled, mainly due to the high costs of the process in the USA, Canada, France, and Norway. Most of the big magnesium plants in those countries were closed due to lower competitiveness. Despite the establishment of duties that could reduce imports, U.S. producers of magnesium began to exit the market. In 1998, Dow Chemical decided to leave the magnesium business, contracting a licensing deal for its technology to Samaj, a Pima Mining's subsidiary, for the South Australian magnesium project. Northwest Alloys Inc. closed its plant in Washington by 2001. Renco Metals Inc., the Magnesium Corporation of America parent, filed for chapter 11 bankruptcy in August 2001. In the same year, Norsk Hydro ended magnesium production in Norway, and after 6 years, in 2007, it ended its

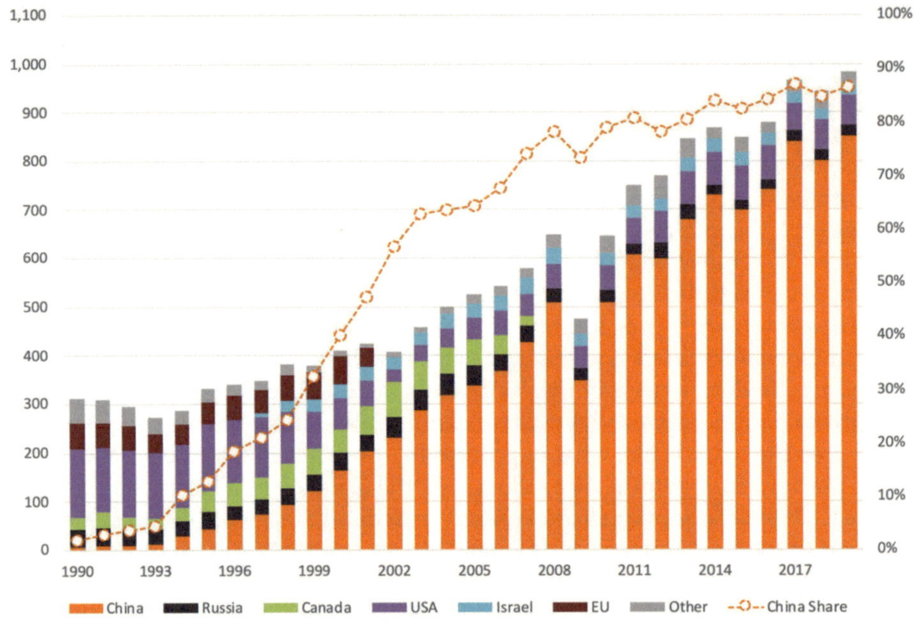

Figure 5.
Evolution of magnesium production per region (1990–2017) [16].

operation at Becancour, Canada. Noranda, which operated in Quebec the Magnolia electrolytic magnesium plant relied on serpentine tails from nearby asbestos mines, closed the smelter in 2003.

By 2015, more than 80% of the world's magnesium production took place in China, followed by Russia, Israel, and Kazakhstan, with only a few percent market share. In 2021, due to curbs in domestic power consumption, Chinese production of magnesium had been halted or curtailed to such an extent that deliveries to Europe have drastically dropped since 20 September 2021. In the second half of 2021, in the world's main magnesium production hubs, Shaanxi and Shanxi Provinces, 25 magnesium plants would have to shut down. Five other plants had to cut production by 50% amid China's power curbing rollout. With an 87% global share in magnesium production, the Chinese supply shortfall has already resulted in record prices, reaching the never recorded price of 6 Eur per kg and a worldwide global distortion in the supply chain.

The dependency on Chinese producers has created magnesium users worldwide a deadly embrace. Fluctuant prices over the 2000s depend on Chinese supplies. From the end of 2007 to the end of the first quarter of 2008, the average U.S. spot Western price increased significantly, as in China and Europe. Several factors contributed to these price escalations. In the United States, a decline in imports from Russia and Canada, two of the leading import sources, caused a supply shortage on the spot market. In China, increased prices for ferrosilicon, power, and transportation were causes for the rapid price increase [1]. In addition, environmental crackdowns by the Government of China may have led to shutdowns at some smaller and highly pollutant Pidgeon plants. In the United States, the Platts Metals Week U.S. spot Western price range reached a peak of USD 3.50 to USD 3.70, while in China, the magnesium price range reached a high of USD 5950 to USD 6250 per metric ton. The increased production cost of Chinese magnesium is firmly attributed to higher prices for raw

	Electrolytic reduction (Western) (%)	Thermal reduction (China Pidgeon, coal-powered) (%)
Materials	6	50
Capital	37	20
Energy	18	8
Labour	16	5
Other	24	17
Total	100	100

Table 2.
Cost shares breakdown of primary magnesium and significant differences existing among the old Western big electrolytic plant and the small Pidgeon plants.

material (main ferrosilicon), decreased production due to stricter environmental regulations at smelters and coal mines, increased labor costs, and an increase in coal power cost. **Table 2** represents the cost-shares breakdown of primary magnesium and significant differences between the old Western big electrolytic plant and the small Pidgeon plants powered by coal, primary actors of national magnesium production expansion in the first decade of the 2000s.

Though the raw material cost is essential, price stability is a much more relevant factor. For this reason, several projects are currently being developed to increase primary magnesium production capacity worldwide. In Nevada, United States, one company has obtained permission to build a pilot plant to test magnesium production from a dolomite deposit. In Quebec, Canada, a company started the construction of a secondary magnesium smelter. A company in Australia with a 3000 ton per year plant is going to be completed; it will recover magnesium from coal fly ash [17].

Now, let us go a bit in-depth about price concerns.

On the one hand, manufacturers are under the constant pressure of product costs that must be affordable; on the other hand, they cannot easily justify the use of bright material characterized by a (historical) uncertainty of supply over a medium-term period. **Figure 6** shows the price history of magnesium metal (US Market spot price) relative to magnesium and aluminum [USGS Bulletins]. On that source, it is crucial to notice that the ratio between magnesium and (primary) aluminum price has been over the ratio of 1.6, which is generally considered the affordable price ratio for magnesium versus aluminum, usually calculated by the inverse ratio densities of the two materials.

On the other hand, it would be more appropriate to consider the switching cost for each kg of steel that you would substitute with the alternative light metal for the same function. **Table 3** represents a viability study on the structural application of light metal alloy for manufacturing the automobile outdoor body panel that shall guarantee equal (or higher) stiffness and denting capability. To evaluate whether it is technically convenient to replace galvanized mild steel with lighter aluminum and magnesium metal alloys for stamping an outer door panel of a road vehicle, we need to know for alternative lightweight scenarios the substitution factors that are defined as the mass ratio between the lightweight (aluminum and magnesium) and the baseline (steel) component. The mass is obtained by multiplying the material density by the volume of the panel. Otherwise, the outer door panel volume is obtained by the front area of the panel that is usually fixed due to geometry constraints (e.g., the perimetral geometry defined by screen and center pillars) and the thickness of the panel sheet cold drawn.

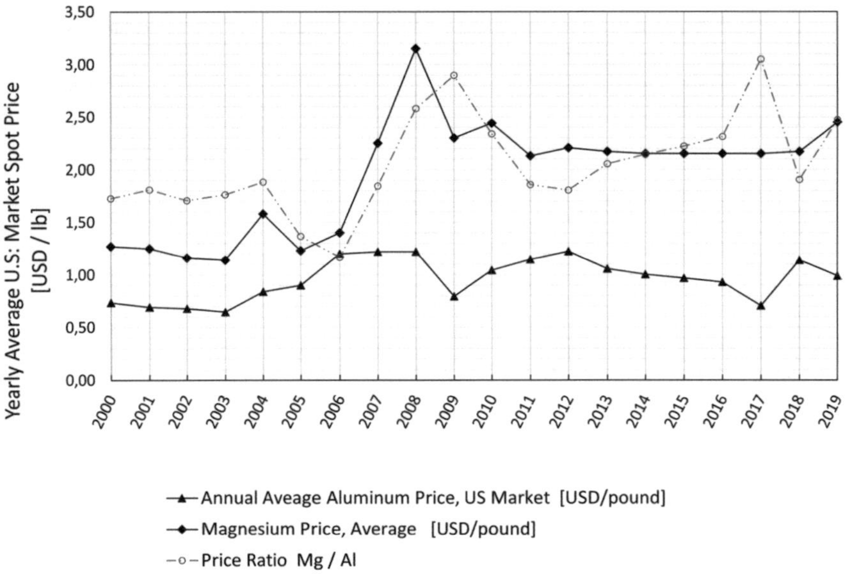

Figure 6.
Yearly average U.S. market spot price for aluminum and magnesium [18].

	Baseline	Lightweight solutions	
	AISI 1045 steel	**Aluminum 6061**	**AZ31D twin rolled cast sheet, warm formed**
Density [g/cm^3]	7.8	2.7	1.8
Yield strength (minimum) [N/mm^2]	350.0	250.0	135.0
Elastic modulus	210.0	70.0	45.0
thickness for bending load parity and stiffness [mm]	0.8	1.2	1.3
Strength-to-weight ratio (kNm/kg)	44.9	91.2	75.0
Minimum thickness to achieve bending load parity with galvanized steel [mm]	—	1.2	1.3
Minimum thickness to achieve stiffness parity to galvanized steel [mm]	—	0.9	1.3
Panel mass [kg]	15.9	8.1	6.2
Substitution factor (SF)	1.0	0.5	0.4
Weight saved in percentage with lightweight solution	0%	49%	61%
Current production cost (Eur/kg)	0.9	3.3	6.5
Elastic module [GPa]	210.0	70.0	45.0
Switching cost per each kg steel substituted with an alternative metal for the same function, equivalent stiffness, and load (denting) capability, calculated as (price of 1 kg material) × SF: [Eur]	0.9	1.7	2.5
Material unitary cost for steel parity, calculated as (unitary steel price)/SF: [Eur/kg]	0.9	1.8	2.3

Table 3.
Feasibility study about the affordability of lightweight solutions with aluminum and magnesium alloy for an outdoor body panel for the automobile; comparison with baseline steel scenario.

Furthermore, it is a multiple constraints problem: it is a strength-limited design problem with constraints in terms of the same (or higher) dent resistance and same (or higher) flexural stiffness of the panel. Under these circumstances, substitution factors for an aluminum alloy AA 5083 sheet cold drawn ranges 0.5–0.6, for a magnesium alloy AZ31D twin rolled cast sheet warm stamped ranges 0.4–0.5 (refer again to **Table 3**).

Thus, by calculating the material substitution factor for each light metal considered, we would evaluate how much is the switching cost of each kg of steel when it is substituted with 0.6 kg aluminum alloy or with 0.4 kg magnesium alloy. **Table 3** shows the switching costs per kg of steel in the case of both aluminum and magnesium solutions. Much more interesting is the line indicating the "steel parity" unitary material price (Euro/kg) for the outdoor panel: it represents how much it should be the unitary price for an alternative material to manufacture the body panel at the exact cost of the baseline case, the steel made pan.

Hence, the big question: is the steel parity cost the unique parameter to consider if magnesium is attractive as light material?

3. Magnesium for weight-saving and lowering energy for transportation

Nowadays, the transportation sector impacts around 25% of direct CO_2 emissions from fossil fuel combustion. Among the variety of transportation means, road vehicles, particularly passenger cars and freight vehicles such as heavy trucks, busses, and two-wheelers, are estimated by the International Energy Agency (IEA), accounting for nearly three-quarters of transport CO_2 emissions. Although CO_2 emissions from aviation and shipping have been increasing in the last decade, the road share of total transport sector emissions has fluctuated around 75% of total transport emissions for two decades. If several efforts and advancements have been made in road-vehicle electrification, otherwise larger (and heavier) vehicles are still preferred by lots of consumers. The worldwide market share of SUVs has increased in the last two decades, and in 2019, before the pandemic crisis, it represented nearly half of the global light-duty vehicle market in several countries. Growing demand for the urban transport of goods is rising, adversely affecting air quality, noise, safety, and liveability in the city. The automotive sector has been putting efforts for reversing CO_2 emissions growth by several strategies; one of those strategies focuses on energy efficiency countermeasures that would be implemented in the form of:

managing/rationalizing travel habits to reduce the frequency and/or distance switching from high-energy-intensity modes (e.g., private car and or air) to most efficient methods (i.e., train for traveling long distances plus rented new efficient vehicles on local base).

deploying energy-efficient technologies for vehicles and fuels.

more stringent requirements on vehicle efficiency, namely, power consumption per km.

The latter strategy is thought a valuable approach for accompanying market migration from heavy vehicles powered by combustion engines fueled by gasoline toward cleaner electrified cars that could be likely powered by near-zero-emission electricity. During vehicle operation, the fuel consumption rate can be approximated as the sum of a linear function of the vehicle mass and—as a second contribution—the loss in aerodynamic drag; both of them through coefficients that depends on several vehicle characteristics. Strategies approaching weight reduction are actually most effective during transient driving cycles; instead, during constant speed traveling, the vehicles' fuel efficiency mostly depends on aerodynamic drag forces. Global average fuel consumption of new cars has been too slowly decreased, less than 2% per year, setting around 7 L gasoline equivalent per 100 km (Lge/100 km). To get on track with 2030 targeted 4.5 Lge/100 km, expected standards will become significantly more stringent to achieve efficiency goals. In 2021 the European Commission proposed new CO_2 emissions targets for 2030 and 2035 that require CO_2 emissions reductions of half actual emissions for cars and vans.

Despite wide literature on life cycle assessment of on-road vehicles considers fossil fuel-powered vehicles, a similar approach is being deployed in the case of electric motor-powered vehicles (considering the energy efficiency of kWh per km traveled) or hydrogen-gas fueled road vehicles (considering hydrogen gas supplied to fuel cell unit per km traveled). Precisely for fully electric cars, the weight of full-electric vehicles is a sum of the mass of the vehicle's architecture and the mass of battery packs. Thus, its common sense considering that the travel range represents for the consumers the independence from the plug-in commences with battery size. That's the Achille's heel of plugged-in vehicles for fossil-fueled vehicle buyers. On the one hand, travel range increases with battery capacity, but on the other hand, larger battery capacity means a heavier vehicle to travel.

While vehicle downsizing improvements in fuel efficiency could be achieved, it appears in contrast with buyers' needs; thus, reducing size for reducing mass could not target a competitive strategy for automakers.

For this reason, a weight-saving strategy primarily implemented by extensive use of lightweight materials—better to say, by higher specific strength—is, therefore, most promising for pursuing consumers' satisfaction. Meanwhile, environmental aspects are successfully addressed, as they cannot be deferrable. It has been estimated that a 10% of vehicle weight curbing increases the vehicle's fuel efficiency by nearly 7%. But the ability to introduce new lightweight materials into vehicles is not a simple remove-and-replace process. Concerns about the impact of material changes on manufacturing lines, supplying network reliability, material cost stability, secure material availability in the marketplace are the main drivers in the material-shift decision process as they all could be more important for automakers than the percentage of weight saved. The potentiality of any lightweight scenarios steered by material replacement rates is based on the actual capability of lighter but weaker materials to safely replace heavier but stronger ferrous alloys, like steels and cast irons. As shown in **Table 3**, the weight-saving potentialities of lighter material depends on the specific substitution factor for the specific function, and it's a fact that the materials substitution factor strongly depends on: the physical properties of the material (e.g., its density and its elastic module as key-factor impacting on stiffness-limited design), the shaped part mechanical properties that are strongly dependent on the shaping process employed (e.g., fatigue limit obtained by cold pressure die-casting operations is different from fatigue limit obtained with low-pressurized die-casting), the geometrical constraints fixed by design (e.g., limited space of fixed boundaries to frame architecture).

As it is usual for any comparative analysis, we need a baseline and parameter to use in the calculation of data output to compare. The fuel consumption reduction coefficient is conventionally used as a measure of fuel-mass correlation. It provides the saving in specific consumption achieved through a 100 kg weight-saving. Recent literature set in the range 0.3–0.5 L/(100 km × 100 kg), varying with modeling assumptions, such as vehicle class, car model, driving cycle, the fuel consumption reduction coefficient for internal combustion engine vehicles [19], and values in the range of 0.47–1.17 kWh/(100 km × 100 kg) for electric vehicles [20].

But there is always a "but"; light-weighting is not a stand-alone measure whether its motivation is pollution curbs [21]. The extractive metallurgy (mining and refining) of nonferrous structural metals that are highly reactive toward oxygen like aluminum, titanium, and magnesium is complicated due to their low grade. The high complexity of the ore extraction and the energy-intensive pyrometallurgical or hydrometallurgical processes employed for pure metal refining are critical stages for the potential release of gas, liquid, and solid emissions (i.e., direct pollutant emissions) and for a large amount of CO_2 emissions correlated to lots of energy consumed (i.e., indirect pollutant emissions). In the next sections, we'll go into details, but for the moment, we can summarize by this way:

- light nonferrous alloys are the cleaner solution in their usage phase in the transport sector because they can target a consistent weight curb capable of reducing direct emissions at the tailpipe of fossil-fueled vehicles. For internal combustion engine vehicles with average tailpipe emissions near 120 gCO_2/km, assuming 2.4 $KgCO_2$ emitted al tailpipe per liter gasoline burned, it can be calculated to reduce by 100 kg onboard vehicle mass, nearly 1.2 $kgCO_2$ is the pollution cut per 100 km traveled.

- The direct CO_2 curb obtained in the usage phase could be reduced (sometimes nullified) by the sum of pollution (direct and indirect) emitted during the extractive phase, the refining process, and the manufacturing of the lighter products. For example, on average, the greenhouse gases emitted as equivalent CO_2 per kg of product manufactured can vary from 2.0 to 2.5 $kgCO_2eq$ per each kg of steel made products and 12.0–16.0 $kgCO_2eq$ per each kg of aluminum products.

Therefore, a broad vision must encompass the net CO_2 emissions over the road vehicle lifespan.

A qualitative scheme representing the green ability of light alloys against heavier metal, such as steel, is depicted in **Figure 7**. The baseline case (1) represents a reference, for example, a body panel made of galvanized plain carbon steel. For the steel-made product, the total CO_2 emitted over the product's lifespan is the sum of the CO_2 (direct and indirect) emitted during the manufacturing stage and the usage phase (traveling). By replacing steel-made products with lighter metal alloy (2), we shall consider more pollutant emissions in the fabricating stage. For this reason, the break-even point T1 versus the baseline scenario (1) could be targeted at the T1 traveled distance. The beneficial effect of weight saving is visible by the gray shaded area from T1 to the expected vehicle lifespan representing the net CO_2 curb by lightweight solution. Case (3) represents the use of much lighter material (due to the reduced slope of the line), but with higher CO_2 emitted in the manufacturing stage as per the higher linear coefficient of the line (3). In this second scenario, the break-even point switches to the higher T2 mileage. The difference between the two shaded areas represents the net CO_2 cut for alternative weight-saving scenarios (2) and (3) compared to the baseline scenario (1).

We refer again to **Table 3**. On average, you may consider 0.5 the aluminum substitution factor applicable in structural engineering applications, thus 100 kg of steel (that accounts for around 230 $kgCO_2$ for the manufacturing stage) would be potentially replaced by approximately 50 kg of aluminum, which accounts for about 700 $kgCO_2eq$ emitted in the manufacturing phase. On the other hand, referring to the schematic representation in **Figure 7**, aluminum onboard would potentially save 50 kg. Now you

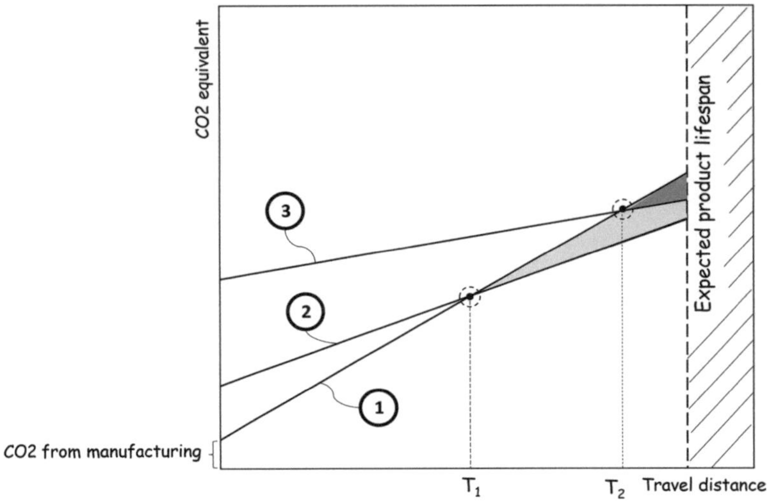

Figure 7.
Three qualitative scenarios for addressing the environmental impact of automotive body panels over the product lifespan.

can account for nearly 0.2 gramCO_2, the pollution cut per kg of weight saved and per each km traveled. Putting onboard an internal combustion engine-powered vehicle 50 kg aluminum to replace steel, we would cut around 5.95 gCO_2 per km traveled.

Now, we can proceed with a further step.

The net CO_2 emissions from the aluminum-steel switch account for around 470 kgCO_2eq emitted in the "cradle-to-gate" phase (including extractive, refining, alloying, and manufacturing stages). Aluminum bodies shall travel onboard around 78,000 km to achieve the break-even point, namely the traveling distance necessary to offset the 470 kgCO_2 extra emissions over the steel-made bodies (the baseline scenario). The environmental sustainability of the lighter solution is therefore strongly influenced by the environmental impact of the raw material fabrication phase, mostly the extractive stage. For that reason, intensive use of recycled patterns to limit the use of primary (virgin) metal for such energy-intensive lightweight alloys is the key to excellent sustainable use of light alloys on-road vehicles. And what about magnesium products? Former data about the carbon footprint of magnesium production have indicated an extensive range of 37–47 kgCO_2eq/kg of magnesium [22]. With such numbers, many still consider magnesium from a technical point of view an exciting opportunity to implement lightweight strategies but an unsound option for a cross-cutting greening approach. It would be effortless to calculate whether 37 kgCO_2eq is the carbon footprint per kg of magnesium to put onboard for replacing 1.6 kg of steel, the CO_2 emitted for the manufacturing phase could be "absorbed", traveling for a lot, above 200,000 km. What are the reasons for such a high carbon footprint of magnesium metallurgy? And shall we consider those numbers still valid today? We'll try to get an answer to those questions in the following sections.

4. Safety and environmental concerns about magnesium

There are two primary sources of direct and indirect pollutant emissions in the magnesium industry. The first source is correlated to high reactivity with the oxygen of magnesium when molten. In the air, molten magnesium is coated with an oxide layer, which, in contrast to the aluminum oxide layer, cannot protect highly reactive magnesium from oxygen. It is true that molten aluminum, too, is highly reactive with oxygen. However, the alumina layer that instantaneously forms on liquid metal in oxygen is dense and not permeable to further oxygen. The magnesium oxide layer formed during melting is characterized by low density. The Pilling–Bedworth number (PBR) explains the high-temperature oxidation behavior of different metals and their oxides [23] by the ratio between the molar volume of oxide and the molar volume of metal. This volume change is responsible for varying types of surface stress in the oxide layer.

For PBR <1, tensile stress on the oxide layer promotes the layer to crack: that's what happens in molten magnesium metal. When the PBR is equal to 1 (the better situation with PBR above 1), it represents a safe condition. Sound compressive stresses develop in the thin dense, and stable oxide layer, protecting the molten metal from the outer atmosphere. This happens for iron, aluminum, titanium, and other metals. Unfortunately, this does not apply to magnesium. Emley [24] found that up to 450°C, magnesium forms a protective magnesium oxide layer, but it becomes porous and non-protective over 450°C. The high reactivity with oxygen causes magnesium to easily ignite and endangers the workers and the production line.

The time to ignition depends on the magnesium alloy composition [23, 25]. The real big problem of flame ignition in magnesium is that the oxidation reaction is highly exothermic. Magnesium oxide, white powder, creates a net release of energy in the form of heat. Magnesium ignited burns with flame at more than 2000°C; thus, no crucible can resist if the flame is not extinguished. Furthermore, magnesium atoms are also capable of reducing water to the highly flammable hydrogen gas by the reaction $Mg(s) + 2H_2O \rightarrow Mg\ (OH)_2(s) + H_2(g)$; meanwhile, hydrogen gas could be easily ignited by the excess heat given by the magnesium reduction reaction. Magnesium metal can also react with carbon dioxide when present in the atmosphere to promote and sustain magnesium oxide formation accordingly with the following: $Mg(s) + CO_2 \rightarrow 2MgO(s) + C(s)$. For this reason, conventional carbon dioxide fire extinguishers cannot be used for extinguishing magnesium fires (required Class D dry chemical fire extinguisher or covering the fire with sand to remove air source).

This hazardous behavior of magnesium metal is therefore historically correlated to conditions that lead to flame ignition of molten magnesium or magnesium in the form of powder, ribbon, thin strips, and foils, namely those fine structure forms that can be quickly heated up just by relatively low heating source, for example by friction. The highly exothermic oxidation reaction could bring explosive hazards in the presence of moisture when flame ignition is not adequately managed by specific knowledge and expertise. For these reasons, particularly in the presence of molten magnesium (for example, during cast shaping), fluorine-based compounds, such as SF6, for protection of molten magnesium have been used since the 1930s [26]. Before introducing SF6, magnesium was protected with alkali metal halide fluxes, sulfur dioxide (SO_2), or even elemental sulfur. The decomposition and following reaction between the fluorine and liquid magnesium keeps separate highly reactive molten magnesium from oxygen. On the one hand, these reactions are thought capable of creating on the molten metal surface an elastic, nonporous protective film containing MgO and MgF_2 with a Pilling-Bedworth ratio larger than 1 [27]. On the other hand, significant impact is ascribable to the use of SF6 as a cover gas. The SF6 environmental impact has been calculated to be 22,800 kg CO_2eq/kg of SF6 used (in other words, 22,800 times greater than 1 kg of CO_2 emitted). Usually, 1 kg of SF6 is required as cover gas per ton of melting magnesium, resulting in a 22,800 kg CO_2 equivalent per kilogram of melt magnesium. In Europe, SF6 is banned, while in the United States, its use is optional for the industry.

To present date, banded SF6 has been substituted by less impacting hydrofluorocarbons such as HFC-134a, however, considered a greenhouse gas but much less impacting. A much lower impact is for sulfur dioxide, but it presents limits for its toxicity and its corrosive properties. Usually, a specific blend of them is used. Recently the Novec 612 fluid—registered by the 3 M Company— promises a meager global warming potential (GWP, expressed as $kgCO_2eq/kg$ product) of 1, equivalent to CO_2. Furthermore, to limit the intense use of protective substances, an old approach recently proposed consists of adding unique alloying elements to improve the ignition resistance of magnesium alloys. In the past, Emley [24] claimed that additions of small amounts of Be, Al, and Ca enhanced the oxidation resistance of solid Mg alloy near the melting point. Such magnesium alloys could be melted in the air if the oxide skin on the ingot was not broken. Sakamoto et al. [28] verified the oxide film on the Mg-Ca consists of a CaO surface thin layer, and just below this layer, a mixture of MgO-CaO exists. To date, the main reason for this protective effect from Ca-O is not clarified. One prominent hypothesis embraces the PBR rule. The higher thermodynamic stability of Ca-O added oxide layer and the kinetics of the diffusion and reaction of Mg ion at and through the oxide layer formed by a mixture of MgO and CaO. When the oxide layer consists of a combination of MgO and CaO, the large volume of CaO might compensate for the shrinkage due to MgO formation. Phenomena involved in retarding flame ignition in Mg alloys systems when alloyed with Ca, Be, and Y has been studied for years but not wholly clarified today.

To summarize, reasonable and sustainable practices are available today in the marketplace to safely treat magnesium and significantly reduce the pollutant emissions in handling molten magnesium in foundries.

But the second source of pollution for the magnesium industry, much more relevant and challenging to control, depends on the vast amount of energy necessary for the magnesium extractive and refining phase, namely the primary magnesium fabrication. There are only a few processes available for the primary magnesium fabrication as they are based on sources of the raw materials by which magnesium can be extracted: raw materials ores (such as dolomite, magnesite, hydroxide mineral brucite, halide mineral carnallite) and brine, which is a mainly a highly concentrated water solution of common salts like hydrated magnesium chloride, magnesium sulfate and magnesium bromide, whose preferred reservoirs are the higher concentrated seawater such as the Great Salt Lake and the Dead Sea. By the way, magnesium raw material sources are considered practically inexhaustible, as magnesium is the 4th abundant metal in the Hearth crust, following iron, aluminum, and silicon. Moreover, inexhaustibility is properly true for seawater reservoirs of magnesium chloride salts. Depending on the type of magnesium source employed, we can distinguish two prominent process patterns to produce magnesium metal: (a) the electrolysis of fused anhydrous magnesium chloride obtained by various refining upstream processes (e.g., dehydration of magnesium chloride brines or chlorination of magnesium oxide) and (b) the thermal reduction of magnesium oxide by ferrosilicon derived from carbonate ores. Today's electrolytic processes are mainly based on the oldest and original Dow process employing seawater as a primary magnesium source.

The Dow process was developed in the first decade of the twentieth century, as the USA started an extensive magnesium production for military scopes. Electrolytic cells are vessels equipped with multiple steel cathodes and graphite anodes partially submerged in the dehydrated molten salt electrolyte. They generally operate to temperatures from 680 to 750°C to develop the basic reaction: $MgCl_2 \rightarrow Mg$ (liquid metal) + Cl_2 (gas). While the Dow process was starting and ramping up US national production, Germans continued investigating carbo-chlorination of magnesite to produce liquid anhydrous magnesium chloride. During those years, when the second peak of magnesium demand rose, Canadian scientist Lloyd Montgomery Pidgeon developed the thermal process for reducing magnesium oxide with silicon in externally heated retorts. Silicon is generally obtained by ferrosilicon ores, and it is produced in an arc furnace, mixed with calcined dolomite, and then briquetted. The briquettes are placed in a retort and heated to extract magnesium vapors condensed at the cold end of the retort with a relatively small diameter. The process is a batch process. It requires metal to be removed from the condensers, slag to be evacuated as a solid, and finally, it is possible to recharge the retort. Thus, the Pidgeon process has reduced the productivity of magnesium per day compared to big electrolytic cells plants. We would simplify the basic reaction as: $2CaO + 2MgO + Si \rightarrow 2Mg + Ca_2SiO_4$. It is an endothermic reaction, and a large amount of heat must be applied to initiate it and continue.

Therefore, the Pidgeon main environmental problem is the combustible used for firing furnace; oil or gas are commonly used for the scope. Former literature ascribed to oil firing the high environmental impact of polluting emissions, ranging from 37 to 47 $kgCO_2eq/kg$ of Mg extracted [22].

Another thermal process, the Bolzano process, like the Pidgeon process, employs the dolomite-ferrosilicon briquettes. Briquettes are stacked on a unique charge support system through which internal electric heating is conducted to the charge. In that case, most carbon emissions are drawn by the indirect carbon emissions of the energy mix used. Depending on the electricity carbon footprint on a local base, the Bolzano process ranges from 13 to 33 $kgCO_2eq/kg$ of Mg depending on the local electricity share produced by hydropower [29].

In brief, we can summarize. On the one hand, the Pidgeon process advantage consists of low investments to recoup, fewer constraints on the minimum size to be profitable, short time for facility construction, equipment installation, and plant ramp-up, flexible production. On the other hand, it suffers from low productivity, high labor requirement, and high energy consumption.

But to reply to the big question: Are the environmental concerns about magnesium extractive processes still valid today? It is necessary to ponder data from the most recent life cycle assessment studies committed to an in-depth investigation of magnesium green ability. In 2013, the International Magnesium Association (IMA) published the study "Life Cycle Assessment (LCA) of Magnesium Components in Vehicle Construction" which analyzed the entire life cycle of magnesium components for transport applications [30]. The study addressed primary magnesium production, alloying, component production, use phase, and the end-of-life of magnesium components, particularly for passenger vehicles comparing differences in emissions among Pidgeon process employed in small factories during the first decade of the 2000s with the most recent Pidgeon process practiced in larger plants. The worst numbers in former LCA studies before 2011 addressed massive emissions from small Pidgeon process plants developed in China from the 1990s to 2011. Pollution emissions from small plants significantly decreased following the imposed shutdown of several high pollutant firms, including small magnesium plants, ordered by the Chinese Government 1 month before starting the Beijing Olympic Games to improve air quality. Small factories in the primary magnesium business would have targeted more stringent environmental prescriptions before they could resume production, but several small factories had not restarted production. Survived small plants restarted, at higher operating costs, improving the energy Efficiency with substitution of coal by gaseous fuels, with more efficient re-use of waste heat, and installing additional air treatment equipment.

Therefore, the 2013 LCA analysis published by the International Magnesium Association downscaled the overall average emissions from the Pidgeon process under the improvements mentioned above to 28 kg CO_2eq, including all upstream processes [30]. It is worth noticing the magnesium production plant located in Brazil uses a silicothermic process, a modified type of the Bolzano Process. It targeted an excellent result of 10.1 kgCO_2eq/kg magnesium.

Alternatively to the Pidgeon process, the big electrolytic plants could have a meager environmental impact, depending on the primary energy source. The Dead Sea Magnesium plant, which produces magnesium from the Dead Sea evaporite deposits in Israel, uses natural gas as an energy supply. The global warming potential of this process is accounted for 17.8 kg CO_2eq/kg Mg [30]. As in this electrolysis plant, two main by-products are produced, liquefied chlorine (Cl_2) and KCl-rich salt. They can have a wide range of potential uses; thus, they are used as raw materials for other sectors. Credits for their re-use, therefore, contribute to decreasing the global warming potential to 14.0 kg CO_2eq/kg magnesium [30].

Since 2017 a new electrolysis plant with a capacity of 100,000 metric tons per annum has been operating in China by the Qinghai Salt Lake Magnesium Co. (QSLM). The QSLM electrolytic magnesium smelter is located at Golmud in Qinghai Province. This process produces pure magnesium from magnesium chloride ($MgCl_2$) brine, an adjacent potash production waste product. The smelter produces low CO_2 embedded magnesium metal thanks to energy power used for the complex supplied from regional hydro facilities (75%), solar (9%), and wind, as well as a local thermal power station. With support from the Qinghai Provincial Government and the national Government in Beijing, QSLM has plans to expand the production of pure magnesium alloys from current levels to 150,000 mtpa soon and then to 450,000 mtpa. Adjacent to the electrolytic magnesium smelter, Magontec has constructed a new primary magnesium alloy cast house facility with an output capacity of 60,000 metric tons per annum that will take pure liquid magnesium from the adjacent smelter. Magontec's plant benefits the QSLM's energy supply of 75% hydro and nearly 10% solar. The overall greenhouse gas emissions of the electrolysis amount to 8.5 kg CO_2eq/kg magnesium. Apart from pure magnesium, the electrolysis of magnesium chloride produces gaseous chlorine. The amount of chlorine produced cannot finally be predicted at this stage of the project, but a chlorine yield of around 2.5 kg per kg of magnesium can be assumed. This by-product is used as feedstock for the nearby PVC plant. Producing 2.5 kg of chlorine usually leads to greenhouse gas emissions of about 3.2 kg CO_2eq [30]. Thus, crediting these emissions, which the magnesium electrolysis has saved, leads to overall emissions of 5.3 kg CO_2eq/kg of magnesium ingot [30]. The Qinghai plant has not reached its total capacity but is still ramp-up.

It is a fact that government policies of the country in which plants operate play an essential role in the environmental impact of magnesium. The national electricity mix used for plant operations, disposal, and recycling routes and the grade of technical solution development drastically reduce electrolytic routes' effects. Under the updated LCA data, the following **Table 4** recalculates the GWP for the body panel case study (refer to **Table 3**). GWP for the three options refers to average updated data published in [30]. Finally, since GWP are expressed per unit mass of material used, it is necessary to consider the actual usage of light material for the specific application.

	Baseline	Lightweight solutions	
	AISI 1045 steel	**Aluminum 6061**	**AZ31D twin rolled cast sheet, warm formed**
Substitution factor (SF)	1.0	0.5	0.4
$KgCO_2eq$ emitted for substituting 1 kg of steel with alternative metals for same function, equivalent stiffness, and load (denting) capability ($kgCO_2eq$ emitted per kg of material) × SF [$kgCO_2eq$]	2.3	4.9[1]	2.0[2]
			10.8[3]
Steel parity GWP calculated as (GWP_steel)/(SF): [$kgCO_2eq/kg$]	2.3	4.7	3.7

[1]*Average value for Hall-Heroult, primary aluminum globally [31].*
[2]*Average value for electrolytic process powered by high share renewable energy (QSLM plant) [30].*
[3]*Average value for Pidgeon process revised [30].*

Table 4.
Recalculated GWP data for comparative scenarios in manufacturing a lightweight outdoor body panel with light metal alloys.

Table 4 estimates the $kgCO_2eq$ emitted by aluminum and magnesium solution for substituting each kg of steel in the inner door panel for the same function, at equivalent (or higher) stiffness, and equivalent (or higher) denting capability. The calculation method follows:

$$\text{kgCO}_2\text{eq emitted per kg of material} \times \text{substitution factor (SF)} \quad (1)$$

The last line of **Table 2** shows the recalculated GWP for aluminum and magnesium light solutions to the "steel parity" calculated as:

$$\text{GWP for steel parity} = \text{GWP of steel / substitution factor (SF)} \quad (2)$$

The (2) represents the GWP of metal alloys give parity to body panel made of steel at equal (or higher, as for magnesium solution) stiffness and load capability.

5. Recycling capability of magnesium alloys

Recycling metals is critical to their overall sustainability. Magnesium retains most of the necessary physical integrity when recycled if it is not contaminated. Remelting and forming of ingots are the main energy requirement for recycling, but generally, secondary production of magnesium ingots requires substantially less energy than primary production [32]. Otherwise, this process's greenhouse gas emission mainly depends on the selected cover gas; generally, a cumulative 3.6 $kgCO_2eq/kg$ of secondary magnesium produced could be released [33].

Today the recycling of magnesium is technologically feasible. Currently, the primary source of magnesium alloy scrap comes from the magnesium die-casting industry (the most common method of fabricating new magnesium alloys parts). Die casting foundries can manage the amount of process scrap in three different ways:

- The scrap can be sold on the open market and downgraded for recycling in other sectors, such as steel desulphurization.

- The scraps can be recycled internally or externally within a closed-loop system; this could optimize the demand for primary raw material saving up 50% primary magnesium in casting automotive parts, rising to 85% for electronic die-cast products. Among factors that influence the number of recycled scraps and recycling ratio optimization we should consider: the amount of material lost in the melting cycle, the number of different components that are cast, the percentage of cast parts that must be rejected during production, the end quality of process scrap, and the recycling operation efficiency all affect the amount of process scrap, and primary magnesium utilized [34].

- Remelting of magnesium chips from machining of die castings, considering that, due to high magnesium susceptibility to oxidize and fine forms of chips, chips remelting could produce further dross quantity. For this reason, preliminary treatments are required to eliminate possible lubricants (e.g., aqueous washing treatment, steam treatment). The second treatment in liquid acid pickling solution (deoxidizer) specific for magnesium alloys helps to reduce oxide contamination.

- Recent studies [35] successfully validated the no-melting route for recycling magnesium chips by hot forging and extrusion as it is done for aluminum chips.

On the other hand, the processing of end-of-life vehicles is today still not easily practicable and needs technological improvement. Shredded magnesium can be contaminated with iron, nickel, and copper from coatings and fasteners, all of which are detrimental to the corrosion resistance of the metal. Although the low-value markets mentioned above (aluminum alloys and steel desulphurization) can absorb low-quality post-consumer magnesium, options for separating it from other metals are necessary for magnesium structural alloys applications. The main viable option is melting magnesium in contact with molten salt to remove oxides from the liquid metal. This process is today not fully capable of separating magnesium from other metals. Another possibility is metal vapor distillation: due to higher vapor pressure and low boiling point of magnesium compared to aluminum and many other metals.

On the other hand, a too high magnesium-aluminum chemical affinity results in poor separation. A much more viable route is collecting and separating magnesium from the high-quality scraps made of aluminum-magnesium alloys: the beverage can stock (e.g., the aluminum 3xxx series alloys employed for the body, the aluminum 5xxxs series used for edge), and the aluminum 5xxxs series today preferred for body panels of automobiles. To separate magnesium, chlorine gas bubbling through the liquid alloyed metal can remove magnesium by reaction, but such a hydrometallurgy process requires large amounts of chlorine and energy. Further recent advancements in magnesium separation from aluminum alloy scraps focus on electrorefining. It is a process in which metals are purified in an electrolytic cell where the anode is the impure metal, and the cathode is a very pure sample of the metal [36]. To summarize, an efficient recycling route for magnesium by automotive scraps is still an issue. Broad approaches based on the design for corrosion-free assembling and easy disassembling of magnesium parts in the multi-material structure are on track.

6. Magnesium solution in the automotive sector: the present

One significant value that manufacturers usually give to magnesium is its excellent die-castability resources, compared with aluminum. It is mainly due to very low viscosity in the molten state and reduced (or absent) die-soldering phenomena with steel mold-die that broadly extend mold-dies lifespan. The high castability is one metallurgy factor that allows die-casters to realize large, thin-walled, and complex casting shapes. It is due to a less costly manufacturing process that would replace steel-made components by assembling numerous steel stamped pieces or heavily reinforced plastic members [37]. As magnesium alloys can be cast with thinner walls than aluminum, the lower elastic modulus of magnesium alloys can be compensated using located ribs of thin wall thickness that allow restoring stiffness at required values. Secondly, the lower latent heat for solidification of magnesium compared to aluminum leads to considerably shorter casting cycle times, compensating for the lower heat conductivity of magnesium. For a comprehensive overview of die-casting processes and recent advancements, you may refer to [38], here in the following, shortly resumed. Two main casting processes are available for magnesium, the pressure-assisted cold, and hot-chamber injection, with an alternative represented by low-pressure die casting. In pressurized injection casting processes, high pressure is exerted after the liquid metal injection to compensate for metal shrinkage and remove as much possible air entrapped during the shot sleeve movement that accelerated to pressurize liquid metal into the die. The metal solidifies at high cooling rates (higher for the cold chamber than the hot chamber), leaving a fine-grained material (more satisfactory for the cold-chamber process) with secondary dendrite arm spacing in the range of 5–10 μm. As it is usual for any metal, particularly for magnesium alloys, grain refinement is one primary strengthening mechanism capable of saving good ductility and though properties, generally lower for the common magnesium-aluminum alloys containing more than 3–4% of aluminum. If, on one hand, aluminum promotes a strengthening mechanism based on a solid solution, on the other hand, an excess of aluminum (it is limited up to 9%) produces an almost continuous secondary phase of aluminum enriched—the magnesium aluminide, $Mg_{17}Al_{12}$. The magnesium aluminide decreases local plastic resources at the alpha-solution grain boundaries, where magnesium aluminide precipitates.

The long tradition of magnesium automotive part die-casting is proper for magnesium manufacturers, as shown in **Figure 8**, where an example of Meridien's timeline for automobile products is summarized. Magnesium die-casting is evolving in Mercedes-Benz automatic transmission cases, from the first seven-speed automatic transmission case developed in 2003 (**Figure 9**) to the current eight-speed transmission case, still manufactured by magnesium alloy.

Figure 10a shows the recent magnesium die-cast liftgate inner for the 2017 Chrysler Pacifica Mini-Van realized by Meridian Company with Fiat-Chrysler Automobiles. The liftgate assembly consists of:

- Magnesium alloy die-cast internal structure, around 80% of the liftgate at the nominal.
- Wall stock of 2.3 mm, 20% with localized ribbing and thick patches.
- Aluminum wiper bracket.
- Two-piece aluminum outer skin.

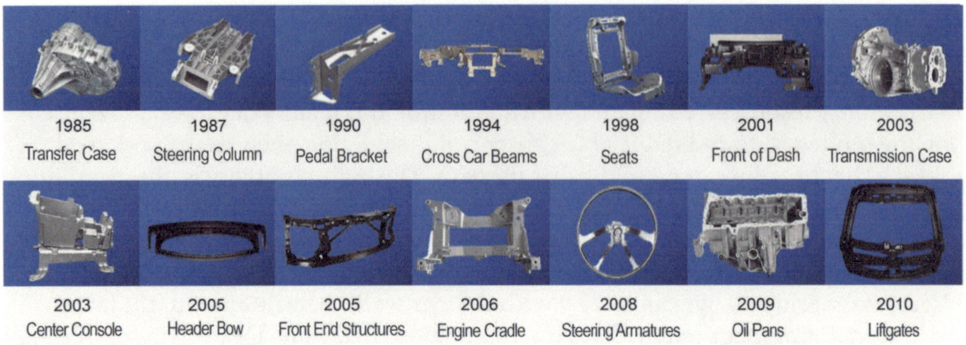

Figure 8.
Meridian product development timeline (courtesy of IMA).

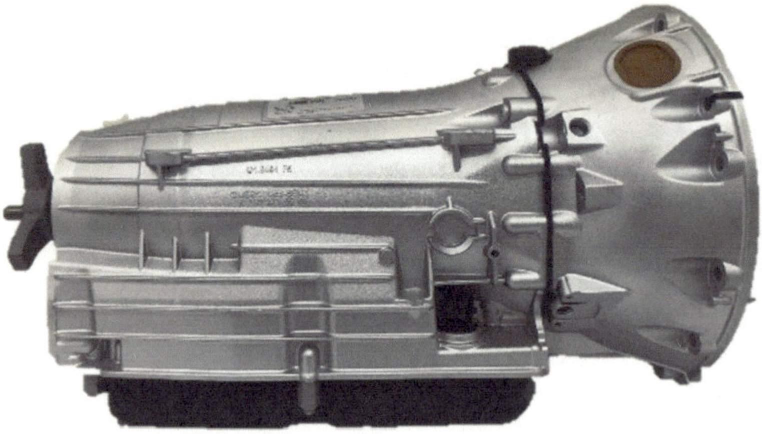

Figure 9.
The 2003 case of the 7G-TRONIC, the world's first seven-speed automatic transmission.

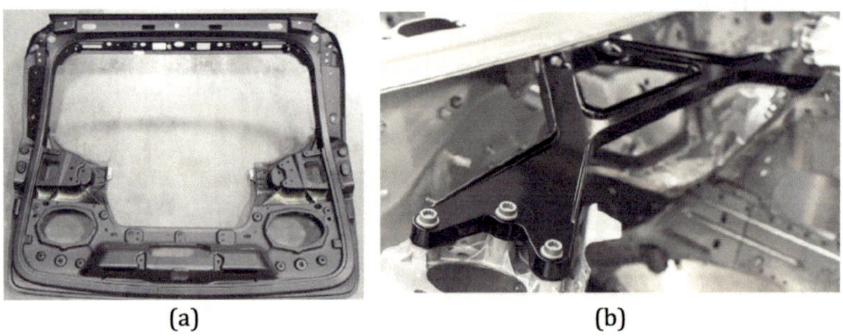

Figure 10.
(a) Diecast liftgate inner by Meridien (courtesy of IMA), 1450 mm wide, 1210 mm in height, the mass of 6.9 kg, (b) Strut bar Audi A8 (courtesy of IMA).

An AM60B alloy has been used due to elongation, strength, castability, and energy-absorbing properties. The magnesium casting allows replacing seven steel stampings, including reinforcements in hinge & latch areas, two plastic pieces, joining technologies. In the final assembly, a powder coat was applied to all structures to prevent galvanic corrosion problems. **Figure 10b** shows the new die-cast strut bar of

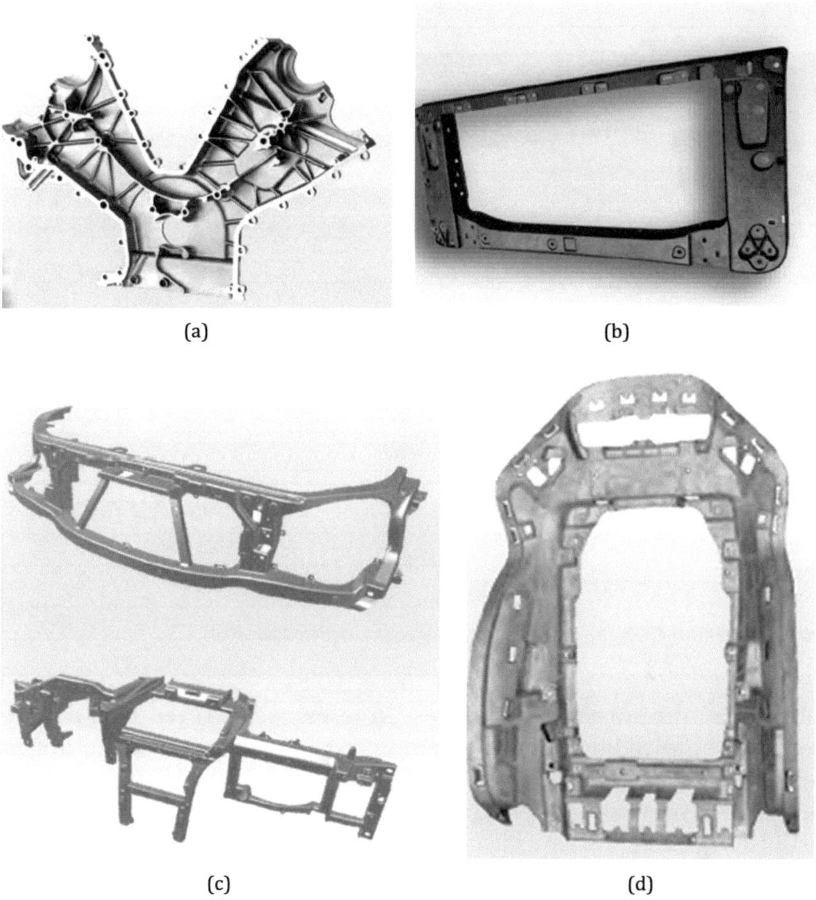

Figure 11.
Recent die cast parts from magnesium industry [courtesy of GF Casting Solutions]: (a) Porsche Control Box Cover, made of MgAl4RE4 alloy, 2.6 kg weight, realized by multistage process (casting, stamping, machining), then assembled; (b) E-vehicle upper door frame alloy made of AM 50 alloy, 2.9 kg weight, realized by multistage process (casting, punching, machining, e-coating); (c) Ranger Rover Front End & Cross Car beam made of AM60 alloy, 6.0 kg weight realized by the multistage process then assembled (casting, machining, stamping); (d) Daimler SLK 2 seat back frame, AM 50 alloy by die-casting, 2.6 kg weight (courtesy of GF Casting Solution AG).

Strut bar for Audi A8 realized by Stihl Magnesium. In **Figure 11**, several magnesium-made parts have been recently put onboard vehicles.

Magnesium part manufacturers deploy a long-tradition cumulated in warm and hot deformation processes (**Figure 12**). Opposite to the common thought about magnesium deformation resources, wrought magnesium alloys are suitable for sheet forming contributing to weight-saving projects in the automotive sector. Indeed, significant drawbacks in magnesium alloys' sheet forming and deformation processes exist, especially compared to aluminum alloys. Due to its hexagonal structure, to activate enough slip systems for assuring sufficient plasticity, magnesium alloys must be formed above 200°C. Furthermore, the different heat transfer capability compared to aluminum is an issue to consider for optimizing pre and re-heating temperatures in the hot-deformation multistage processes. Extrusion of magnesium alloys is usually carried out in multiple steps, starting with a pre-extrusion of large billets into smaller diameter billets. After the preliminary stage, the billet can be re-heated and subsequently extruded into the final shape. Generally, the pressure per unit volume material

Figure 12.
The Chevrolet Corvette SS Race Car, 1957, made of magnesium-formed panels.

extruded is higher than in aluminum alloys. Thus, extrusion speed shall be carefully controlled and optimized for specific magnesium alloy to avoid local melting and local oxidation phenomena, particularly critical for hollow sections extrusion process by porthole dies, as typically employed for aluminum alloys. These aspects are firm limits for semi-finishing and net-shape forming processes, prominently for affordability.

But on the other hand, warm deep drawing is also possible for magnesium alloys, as for aluminum alloys; in the range of 100–180°C thin sheet of 0.5 mm approximately can be drawn, with precise temperature control and at a lower speed [39]. The recent application of magnesium sheets we can find in the literature illustrates the successful use of a novel Mg-Zn-Ca-Zr alloy in sheet form produced by twin-roll casting. This alloy has been used to manufacture a Volkswagen Passat decklid magnesium-made that saved half 6 kg of the original 12 kg steel weight [40]. Large magnesium components can also be produced by die casting (see **Figure 13**).

Net-shape semi-solid forming has attracted automakers with alternate attention since the middle 1990s. The net-shape semi-solid forming is possible for magnesium alloys thanks to its thixotropic state realized when vigorously sheared in a semi-solid state. Shearing reduces the viscosity of the slurry mass to a similar value of the liquid metal, providing similar (sometimes better) castability of the liquid form. Still, the semi-solid state allows shaping with lower latent heat in the mass; this creates advantages for shorter casting cycles than die-casting (depending on chosen semisolid process) lower metal shrinkage to compensate, and consequently near-net shaping. The industrial application of semi-solid net-shaping in the magnesium industry commenced in the early 1980s with the Dow Chemical Company. Dow Chemical patented the Thixomolding technology based on the architecture of plastic injection molding machinery for injecting magnesium alloys in the semi-solid state into a mold die.

Further advantages of injection molding of magnesium alloys are that this technology's highly complex shaping capability allows for more innovative design concepts and a multi-body-material concept design. Direct assembly of different

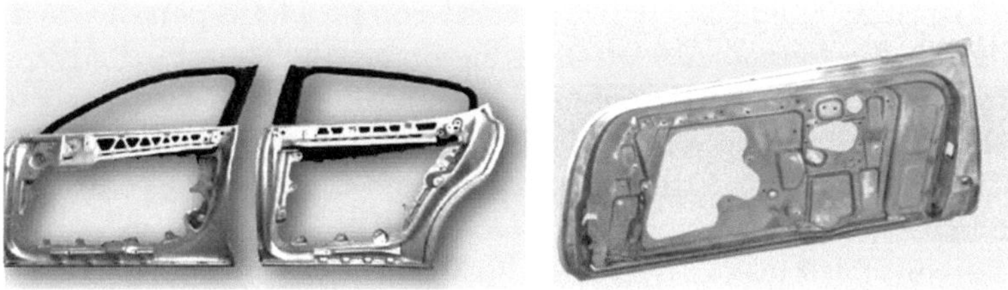

(a) (b)

Figure 13.
(a) The Porsche window frame realized by AM50 magnesium alloy with multi process stages (casting, laser cutting, machining), finally coated and joined; (b) The Aston Martin cover door made AM50 with multi-stage process (Casting, Stamping, Flattening) [courtesy of GF Casting Solutions AG].

parts during injection molding in a molded-in technique, thanks to inserting aluminum parts directly in the tool. As a semi-solid process, less energy is consumed by the Thixomolding apparatus; the power energy is estimated to be on the order of 12–24% lower than the total energy required by a conventional casting process. An additional benefit is that the Thixomolding product cycle employs inert gas, usually argon, to protect magnesium feedstock from oxidation once introduced in the hopper in particulate and solid forms (pellets or chips) [41]. However, it is worth noticing that current die-casting processes align with the environmental sustainability of the Thixomolding process thanks to much more environmentally friendly cover gases mixtures today used instead of the banded SF6. With relevant advantages of the Thixomolding process in net-shaping part of high complexity in a single shot, two are the most drawbacks acknowledged by part manufacturers. The high price and the limited number of suppliers of chip or pellet forms of magnesium alloys, namely the material feedstock of Thixomolding machinery, and the maximum clamping force exerted during the metal injection into mold dies. Clamping forces of 2700–6500 kN generally allow the production of thin flat surfaces (0.8 mm, not possible by high pressure die-casting) such as that for tablet terminals, notebook computers, electronics, sports goods. Instead, the interest of the automotive sector is to even thicker and wider structural components with a weight of over 2 kg. This would require more giant machines with increased clamping forces over 8000 kN [41].

7. Conclusions: new trends and perspectives

The historical and current primary market for structural applications of magnesium alloys is high-pressure die-cast parts. We find those components in the automobile's powertrain, chassis, or body areas. Depending on the type of structural part to shape, key technical features that need to target are the safe-at-break behavior avoiding fast-fracture failure modes, sufficient toughness (i.e., minimum impact energy to rupture and fracture toughness), specific strength, corrosion resistance, high-temperature resistance, or creep resistance (for powertrain applications). From the manufacturing point of view, the requirements addressed shall target affordable production cost, which merges fixed and variable costs, derived from the accounting of investment costs (machinery and tools, energy, labor, etc.) and operating costs (raw material cost, trimming, machining, coating costs, etc.) to recoup.

Specifically, in automotive assemblies, corrosion concerns are crucial. Today high purity versions of magnesium alloys show corrosion properties comparable to aluminum die casting alloys, but galvanic corrosion problems persist when magnesium parts have to be assembled with different materials. Therefore, advancements in coating techniques are the basis for safely combining magnesium parts with other materials. High ductility magnesium alloys are of interest to the automotive sector. Advancements in alloying are crucial for the correct choice of the structural ability of magnesium material. The higher creep resistance of Mg-Al-Si, the AS series, acknowledged by Germans during WII, is allowed by Si addition which forms fine and hard Mg2Si particles along the grain boundaries to help retard grain boundary sliding. The remarkable grain-refining ability of zirconium in the Mg-Zn-Zr series alloys allows high strength and ductility for use either at elevated temperatures or for energy-absorbing applications, however at a higher cost due to Zr. The ZE series achieved further mechanical properties in the die-cast part with Mg-Rare Earth-Zn-Zr casting alloys. RE elements (La, Ce, etc.) are added as they are active during aging treatment by promoting high-temperature stable precipitates with a strengthening effect. The costly Magnesium-Yttrium casting alloys, the WE series, containing approximately 4–5 wt.% Y, exhibit high strength with good creep resistance at temperatures up to 300°C and superior corrosion resistance (comparable to some aluminum-based casting alloys). Furthermore, the WE43 and the Elektron 21, a proprietary magnesium-based casting alloy containing neodymium, gadolinium, and zinc developed by Magnesium Elektron (today part of Luxfer MEL Technology), passed stringent flammability tests of Federal Aviation Administration FAA-FAR 25.853 Part 25, Appendix F, Part 2 Modified Seat Cushion Test. Both alloys did not burn when melted, or they are self-extinguished.

More cost-efficient production routes for sheet products are believed to create new opportunities for the automotive market segment. Considerable efforts have been directed at innovative developments of global efforts in expanding the manufacturing capabilities of magnesium sheets through the twin-roll casting process route, offering many benefits, including a reduction in the number of processing steps and energy savings [42].

Finally, last but not least, die-casting and semisolid process design strategies are similar to those employed for injection molding of structural plastics. But, in general, plastic designs require thicker sections than magnesium die-castings. Both materials allow complex shapes with ribs to enhance stiffness (**Figure 14**), but magnesium die-castings need more minor material for these features than plastics. As a result, magnesium die-castings can be designed more efficiently (less volume, less weight, more significant feature variation) and offer a higher degree of definition than comparable plastic designs, superior mechanical properties, and the capability to integrate several functional design features, material recyclability. The latter feature is not of minor importance, being automakers sensitive to recyclability resources of material used for car manufacturing. Although the material price-based approach leads to the obvious choice of plastics, complex and large parts could present unforeseen internal costs to the product line, negatively impacting final product marketability. A whole approach cost also considers the impact on a company's internal costs structure and the value-added needs of the next customer in the product chain, up to and including the end-user. In a total system cost strategy, the benefits of using magnesium tend to outweigh the lowest material price strategy. This is typical for products like instrument panel structures. The benefits of higher stiffness, elongation, toughness, and design flexibility allow the magnesium part to readily integrate many features in a one-piece to be fully assembled and quickly installed into the vehicle with weight-saving up to 50% compared to plastic designs.

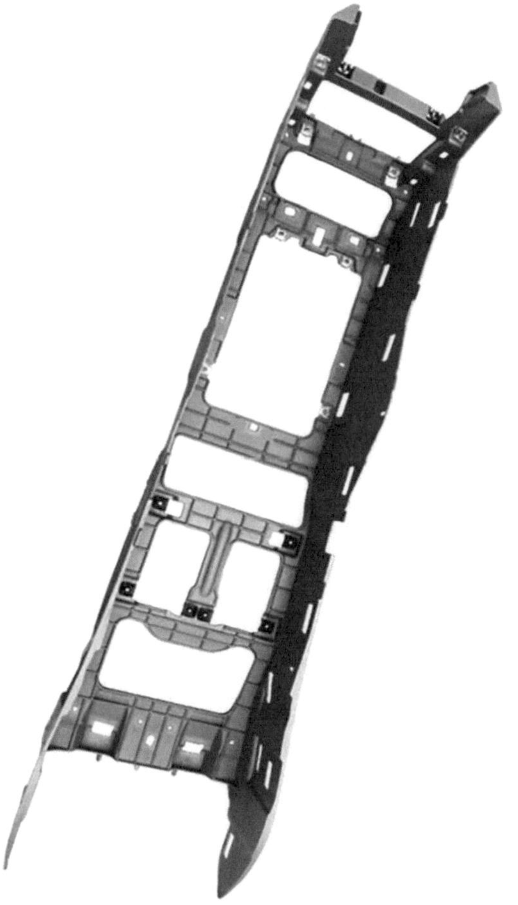

Figure 14.
Magnesium AM 50 die cast front center console for Audi A8, high stiffness, no machining and all connection and fixing points are intergrated (courtesy of GF Casting Solutions).

An interesting overlap of cast magnesium's mechanical and physical properties with reinforced plastics, primarily strength, and density, would drive the material switch. In the interior design of automotive vehicles today, large bodies are made of non-fully recyclable plastics. Thus, other potentialities for magnesium die-casting and injection molding could be redesigning today's plastic structural components with recyclable and more robust magnesium metal.

In this chapter, we tried to resume the magnesium for lightweight approach over the past, till today. Hopefully, but not exhaustively, this was tentative to answer where the magnesium industry is going. We must not forget the past, learning lessons that are still valid today. However, we must consider some new factors, mainly based on the magnesium trade, were unknown in the past century or during the golden Age of magnesium peak demand. It is a fact that when going through magnesium history, several articles projected an optimistic future for magnesium.

Forecasting the future of magnesium, especially in current pandemic times, is difficult. Nevertheless, one aspect appears clear by going through past and recent magnesium history: magnesium had survived continuous fluctuating demand;

meanwhile, price volatility registered over time depended on the current (nonstructured and programmed) supply capacity over time and trade issues.

Several concerns about magnesium's potential applications are today derived from false myths. Eighty years ago, Germans and (after) Americans employed magnesium for aircraft weight-saving, but today it is wrongly thought there are few proofs of its capabilities in realizing lightweight bodies. What is clear from the lesson learned in the past (and today) is that it is necessary to dramatically increase the primary magnesium supply with modern low impacting big plants. Looking at recent history, we are probably on the right track. As learned from the past, prices are not volatile once the supply is stable, and the magnesium's demand (driven by automakers primarily) rises.

Acknowledgements

The authors appreciated the input and feedback by members of the International Magnesium Association (IMA) founded in 1943 with the mission of promoting the use of magnesium in material selection and to encourage innovative applications.

Author details

Fabrizio D'Errico[1*], Martin Tauber[2] and Michael Just[3]

1 Polytechnics of Milan, Department of Mechanical Engineering, Milan, Italy

2 International Magnesium Association, European Representative, Haacht, Belgium

3 GF Casting Solutions, Schaffhausen, Switzerland

*Address all correspondence to: fabrizio.derrico@polimi.it

References

[1] Kramer DA. Introduction and History. Magnesium, Its Alloys and Compounds. Virginia, USA: U.S. Geological Survey; 2001. p. 3. Available from: https://pubs.usgs.gov/of/2001/of01-341/

[2] Bertilorenzi M. Business, finance, and politics: The rise and fall of international aluminium cartels, 1914-45. Business History. 2014;**56**(2):236-269. DOI: 10.1080/00076791.2013.771337

[3] Brandt EN. We Called it Mag-Nificent: Dow Chemical and Magnesium, 1916-1998. Michigan, USA: MSU Press; 2013

[4] Hanawalt J, Glaza G. New applications and developments of magnesium alloys in the automotive industry. In: SAE Technical Paper 520090. 1952. DOI: 10.4271/520090

[5] Patent US2660769A. Dow Chemical Co.; 1950

[6] Wheeler R. Magnesium: What's wrong with it and what to do about it. Light Metal Age. 1966;**6**:17-18

[7] Erickson S. Magnesium: A proven material for light weight automotive die castings. In: SAE Technical Paper 770323. 1977. DOI: 10.4271/770323

[8] Mezoff JG. Magnesium for automobiles. In: Perspective. SAE Transactions. Vol. 89. Pennsylvania, USA: SAE International; 1980. pp. 1617-1630. Available from: http://www.jstor.org/stable/44633783

[9] Hollrigl-Rosta F, Just E, Kohler J, Melzer HJ. Magnesium in the Volkswagen. Light Metal Age. 1980;**8**:22-29

[10] Fisher PA. Desulphurizing iron with magnesium. Light Metal Age. 1973;**6**:19-20

[11] Das SK, Chang CS, Raybould D. High performance magnesium alloys by rapid solidification. Light Metal Age. 1986;**12**:5-8

[12] Mikucki B, Shook S, Mercer W, Green W. Magnesium matrix composites at Dow: Status update. Light Metal Age. 1986;**10**:16-20

[13] Nehan M, Maloney R. Magnesium AM60B instrument panel structure for crashworthiness FMVSS 204 and 208 compliance. SAE Paper No. 960419; Warrendale, PA. 1996

[14] Patent WO 9009251A1. Dow Chemical Co.; 1990

[15] Lieberman MB. The magnesium industry in transition. Review of Industrial Organization. 2001;**19**(1):71-79

[16] CM Group. The Magnesium Report 2017. 2018

[17] U.S. Geological Survey. Mineral Commodity Summaries, January 2020 U.S. Available from: https://pubs.usgs.gov/periodicals/mcs2020/mcs2020.pdf

[18] USGS. Magnesium Statistics and Information, Annual Publications (2001-2020). Available from: https://www.usgs.gov/centers/national-minerals-information-center/magnesium-statistics-and-information

[19] Fontaras G, Zacharof NG, Ciuffo B. Fuel consumption and CO2 emissions from passenger cars in Europe—Laboratory versus real-world emissions. Progress in Energy and Combustion Science. 2017;**60**:97-131. DOI: 10.1016/j.pecs.2016.12.004

[20] Del Pero F, Berzi L, Antonacci A, Delogu M. Automotive lightweight

design: Simulation modeling of mass-related consumption for electric vehicles. Machines. 2020;**8**:51. DOI: 10.3390/machines8030051

[21] Shen D, Phipps A, Keoleian G, Messick R. Life-cycle assessment of a powertrain structural component: Diecast aluminum vs. hypothetical thixomolded® magnesium. In: SAE Technical Paper 1999-01-0016. 1999. DOI: 10.4271/1999-01-0016

[22] Ramakrishnan S, Koltun P. Global warming impact of the magnesium produced in China using the Pidgeon process. Resources, Conservation and Recycling. 2004;**42**(1):49-64. DOI: 10.1016/j.resconrec.2004.02.003

[23] Pilling NB, Bedworth RE. The oxidation of metals at high temperatures. Journal of the Institute of Metals. 1923;**29**:529-591

[24] Emley EF. Principles of Magnesium Technology. Oxford, New York: Pergamon Press; 1966

[25] Kim YM, Yim CD, Kim HS, You BS. Key factor influencing the ignition resistance of magnesium alloys at elevated temperatures. Scripta Materialia. 2011;**65**(11):958-961. DOI: 10.1016/j.scriptamat.2011.08.019

[26] Patent US1972317A. Dow Chemical Co.; 1934

[27] Czerwinski F. Oxidation characteristics of magnesium alloys. JOM. 2012;**64**(12):1477-1483

[28] Sakamoto M, Akiyama S, Ogi K. Suppression of ignition and burning of molten Mg alloys by Ca bearing stable oxide film. Journal of Materials Science Letters. 1997;**16**:1048-1050. DOI: 10.1023/A:1018526708423

[29] Cherubini F, Raugei M, Ulgiati S. LCA of magnesium production: Technological overview and worldwide estimation of environmental burdens. Resources, Conservation and Recycling. 2008;**52**(8-9):1093-1100. DOI: 10.1016/j.resconrec.2008.05.001

[30] Ehrenberger S. Carbon footprint of magnesium production and its use in transport applications. Update of the IMA Report "Life Cycle Assessment of Magnesium Components in Vehicle Construction (2013)". 2020. Available from: https://www.intlmag.org/resource/resmgr/sustainability/2020-LCA-Study-2021-02-09.pdf

[31] Nunez P, Jones S. Cradle to gate: Life cycle impact of primary aluminum production. International Journal of Life Cycle Assessment. 2016;**21**:1594-1604

[32] Sivertsen LK, Haagensen JO, Albright DA. Review of the life cycle environmental performance of automotive magnesium. SAE Transactions. 2003;**112**:275-281

[33] Tharumarajah A, Koltun P. Is there an environmental advantage of using magnesium components for light-weighting cars? Journal of Cleaner Production. 2007;**15**(11-12):1007-1013. DOI: 10.1016/j.jclepro.2006.05.022

[34] Bell N, Waugh R, Parker D. Magnesium recycling in the EU. IMA Report 2017. Available from: http://www.intlmag.org/resource/resmgr/sustainability/FullRprt_EU-Mg-recycling_201.pdf

[35] D'Errico F, Plaza GG, Giger F, Kim SK. Life cycle assessment of ECO-magnesium® alloy produced by green metallurgy EU project process route. In: Alderman M, Manuel MV, Hort N, Neelameggham NR, editors. Magnesium Technology. NY, USA: Springer; 2014, 2014. pp. 7-12. DOI: 10.1007/978-3-319-48231-6_3

[36] Patent US 20150225864 A1. Phinix, LLC; 2015

[37] Ruden T. Replacing plastics with magnesium for structural automotive components. Light Metal Age. 2005;**4**:36-41

[38] Luo AA. Magnesium casting technology for structural applications. Journal of Magnesium and Alloys. 2013;**1**(1):2-22. DOI: 10.1016/j.jma.2013.02.002

[39] Doege E, Dröder K. Sheet metal forming of magnesium wrought alloys—Formability and process technology. Journal of Materials Processing Technology. 2001;**115**(1):14-19. DOI: 10.1016/S0924-0136(01)00760-9

[40] Hernandez JV, Yi S, Letzig D, Kim S, Kim JJ, Seo MH, et al. Magnesium process and alloy development for applications in the automotive industry. In: Joshi V, Jordon J, Orlov D, Neelameggham N, editors. Magnesium Technology. NY, USA: Springer; 2019. DOI: 10.1007/978-3-030-05789-3_3

[41] Czerwinski. Magnesium Injection Molding. Boston, MA: Springer US; 2008. DOI: 10.1007/978-0-387-72528-4

[42] Javaid A, Czerwinski F. Progress in twin roll casting of magnesium alloys: A review. Journal of Magnesium and Alloys. 2021;**9**(2):362-391. DOI: 10.1016/j.jma.2020.10.003

Section 2

Experimental Approaches for Material Applications

Chapter 3

Magnesium in Synthesis of Porous and Biofunctionalized Metallic Materials

Mariana Correa Rossi, Liliana Romero Resendiz and Vicente Amigó Borrás

Abstract

Magnesium particles are used in metallurgic routes, where it can be total or partially evaporated creating pores for ingrowth bone tissue. This book chapter contains the latest findings on the microstructural physical and mechanical properties of β-Ti alloys with Mg additions designed and obtained by the authors. As well as the main new techniques used to fabricate Ti-Mg alloys. An especial emphasis on the microstructure-properties relationship was made to assist on the guide for future efforts of the scientific community towards developing more efficient biomaterials. The β % were related to the low elastic modulus which were in the range of 31–49 GPa close to cortical bone and hardness close to commercial Ti grade 2. The compressive strength was greater than the value of cortical bone. Pore size were in the range of 5–100 μm depending on the sintering temperature, with higher wettability the samples with more porosity. These findings were promising to application of β titanium alloys containing Mg for orthopedic application.

Keywords: oxide magnesium, biological properties, titanium surface, porous metals

1. Introduction

Surgical treatments of bone injuries patients in emergency departments worldwide each year due to involvement in rigorous athletic activities, social instability, traffic accidents, and prolonged human lifespan [1].

Bone defects, mainly induced by traumatic avulsions, sequelae of infection-induced bone sequestration, congenital malformations, or neoplastic resections, confront us with an extreme challenge for reconstructive surgery the need to induce bone regeneration to repair structural bone deficient [2] has inspired research on and development of a vast number of bone repair materials.

Diverse metallic materials are already established as biomaterials due to their high biocompatibility, low toxicity, and good strength–ductility relationship. Examples of these alloys are stainless steel (especially 316 L), cobalt and chromium (CoCr) alloys, and titanium (Ti) alloys [3, 4]. However, the low toxicity and mechanical properties of Ti alloys, specifically the elastic modulus, are more adequate for biomedical uses.

From Ti alloys, the most common for dental and orthopedic applications are materials formed by Ti, aluminum (Al) and vanadium (V) like Ti-6Al-4 V and other β-phase-type alloys as the ones with high contents of β-stabilizers (V, Cr, molybdenum (Mo), Fe, niobium (Nb), and tantalum (Ta)) [5–12]. However, several reports point to the V in the Ti-6Al-4 V as toxic [13, 14], being a motivation for exploring further V-free options. Moreover, the β-phase type Ti alloys have a good combination of mechanical properties and biocompatibility. Besides, the β-Ti alloys have a lower elastic modulus compared to other Ti alloys [15]. Considering that the elastic modulus is a key factor for the success or failure of the implant, this is a remarkable characteristic of these alloys [1]. However, the reported elastic modulus for β-Ti alloys ranges from 69 to 110 GPa [15, 16], being still far from that of human bone (lower than 30 GPa) [17].

To overcome this drawback, several Ti alloys are being developed and most of them are showing promising results in the matter of mechanical properties. A number of these metallic systems are being obtained through powder metallurgy methods to obtain functional porous structures. It has been widely reported that the porous surfaces assist on the fixations and ingrowth of organic tissue, improve the body fluid, reduce the mechanical mismatch due to lower elastic modulus values, and reduce the failure rate of implants [3]. Examples of the above are Ti and indium (In) as (Ti-In) [18], Ti-Mo [7–9], Ti, Nb and Tin (Sn) as Ti-Nb-Sn [10, 19], Ti and zirconium (Zr) as Ti-Zr [20], and Ti and silver (Ag) as Ti-Ag [21] alloys. However, some of the previous systems employ alloying elements that are still not widely studied, being a reason why several *in vivo* tests of biocompatibility should be carried out to determine their biological feasibility.

Another route is the design and development of biomaterials based on widely explore elements as magnesium (Mg). This element has multiple advantages for biomaterials as non-toxicity, biocompatibility, biodegradability, increase strength of the bone, and has a low elastic modulus [3, 4, 22–24]. Low concentrations of Mg^{2+} play an important role in cells activity by stimulating the improvement of cell adhesion and extracellular matrix mineralization [25, 26]. Furthermore, Mg is the fourth most abundant element in the human body and is essential in digestion processes [22, 24]. The non-harmful degradation of an Mg, zinc (Zn) and manganese (Mn) as $Mg_2Zn_{0.2}Mn$ alloy inside the human body has been demonstrated [23]. Based on the above, Mg is a feasible alloying element to boost the biocompatibility and possible control of biodegradability over the time of different biomaterials for medical purposes. The biodegradability of Mg can avoid the need for a second surgical process to remove the implant. The possibility to control such biodegradability is still under intense investigation [23, 27, 28]. Moreover, Mg is a potential alloying element to significantly reduce the elastic modulus. This could reduce the failure rate due to mechanical mismatch between the implant and the bone, and the occurrence of load shielding (absorption of mechanical stress by the implant) [3]. However, one of the main disadvantages of Mg as a biomaterial is that the degradation rate can be faster than the required to allow a complete regeneration of the organic tissue [29]. This is the motivation to explore the use of Mg as an alloying element instead of a matrix. Considering the already explained qualities of Ti biomaterials, it is a good candidate to join with the virtues of Mg.

Until now, few reports on Mg as an alloying element of Ti alloys have been reported [30–33]. Deep research is still needed in the matter of optimizing Mg contents, processing parameters, and designing new systems that reduce the economic and health losses due to the failure of implants. The field of Ti-Mg alloys is emerging and is pointing as highly promising for biomedical purposes.

This book chapter contains the latest findings on the microstructural, mechanical, and biological properties of Ti alloys with Mg additions designed and obtained by the authors. Also, a description the most important techniques to obtain Ti-Mg alloys for biomedical application. An especial emphasis on the microstructure-properties relationship was made to assist on the guide for future efforts of the scientific community towards developing more efficient biomaterials.

2. Chemical and biological properties of magnesium

As the largest dynamic biological tissue in the body, bones are composed of inorganic minerals as magnesium and metabolically active cells surrounded by a large volume of extra cellular matrix, and they form a rigid scaffold that has an irreplaceable role in maintaining life activities, including supporting the body and protecting visceral organs [1]. For bone repair, metallic materials are used to repair or replace the bone tissue damaged. The main materials used in orthopedics include stainless steel and Ti alloys because they are mechanically strong and resistant to fracture. However, there is a potential for the release of metallic ions and/or particles through corrosion and/or wear that trigger inflammatory responses that can reduce biocompatibility and lead to tissue loss [34].

Previously *in vivo* and *in vitro* studies have shown that metallic biomaterials fabricated with Mg exhibit good biocompatibility and free of systemic inflammation reaction or affection of the cellular blood composition. In addition, high mineral apposition rates and increased bone mass were found around degrading Mg implants in bone [35] also, it can present a benefic influence in metallic materials once the bone-implant interface mineralization and osseointegration are significantly greater for metallic materials like titanium and magnesium alloys (Ti-Mg), hence have shown to stimulate new bone formation when used as bone fixture.

Mg is the fourth most plentiful cation in the human body, and is an element essential in many metabolic processes, involved in the regulation of eukaryotic cell proliferation, structural functions are correlates with the enhancement of protein synthesis. Furthermore, is primarily stored in bone tissue, controlling growth of bone cells and accelerates the bone healing [36, 37], which has characteristics of bio-degradability, in the physiological environment can be eliminated and also the corrosion product of Ti-Mg alloys (Mg^{2+}) does not cause unexpected complications because excessive Mg^{2+} are easily eliminated in the urine. Moreover, alloys fabricated using Mg elemental can present mechanical properties similar to those of bone, due to fabrication pores material, decreasing the elastic modulus [38, 39]. Once the bone resorption around stress-shielded are in bone fixation treatments is an important consideration within clinical sectors. Because of its versatility, metallic biomaterials based on Mg can contribute to biological properties and improve the osseointegration process [40].

Mg^{2+} is distributed in three major compartments of the body: ~65% in the mineral phase of bone, 34% in muscle, ~1% in plasma and interstitial fluid [41] and it has a radius of 0.65$\dot{A}$ (Mg^{2+}). Usually, at the physiological pH range, Mg^{2+} is hydroxylated with six H_2O molecules with a large hydration energy to form a complex with a large radius of approximately 5$\dot{A}$, also strongly interacts with phosphate ligands from nucleotides such as ATP and DNA due to its high charge/radius ratio [42]. Coordination number and spatial distribution of water molecules surrounding the Mg^{2+} influence its binding thermodynamics with the protein. As it can be associated with water or phosphate complex binds protein with three to five coordinating oxygen plays a major role in the protein-binding ligands.

Its charge density is approximately $(.99/.65)^3$ 3x more than that of ion calcium (Ca^{2+}) and its affinity for electronegative ligands, almost always oxygen in biological systems, is much greater. Further, the Mg-oxygen bond length is approximately 2.05Å . When Mg elemental is octahedrally coordinated by six oxygen atoms, the oxygen-oxygen distance is $2.05 \times 2^{0.5} = 2.9$Å, an optimal van der Waals contact [43]. In biological systems, Mg ions exists in 3 different states: bound to proteins, complexed to anions, and free (Mg^{2+}). Only free Mg has biological activity [44]. The adult human body contains approximately 24 g (1 mol) of Mg in cells × 280 mg in extracellular fluids and the skeleton represents the body's largest Mg store (approx. 60% of total Mg), divided into two subcompartments. Thus, bone functions as a large Mg reservoir, helping to stabilize its concentration in serum. About one fourth of total Mg^{2+} is located in skeleton muscle [45]. Bone is continuously under load, which can cause bone defects such as fatigue cracks. Such defects can be dealt with by the body's, own healing mechanism. Thus, such critical size defects (CSDs) have to be treated with an implant.

3. Most promising techniques to fabricate Ti-Mg alloys

Ti alloys can be fabricated using as alloying element Mg to create porosity, leading to formation of biocompatible scaffolds with lower elastic modulus by metallurgical and additive manufacturing process. Thus, the stiffness of Mg-based implants can be more easily tailored to match that of bone, which reduces the risk of stress shielding, a phenomenon that will be discussed in the next section.

3.1 Fabrication via powder metallurgy

One of the most conventional techniques to create porous metallic materials is by the powder metallurgy. Conventional manufacturing of Ti alloys via powder metallurgy involves: (1) blending of the powder to achieve a uniform particle size distribution and, if needed, mixing of the Ti powder with the required alloying elements; (2) shaping of the powder blend (this can be achieved via different manufacturing methods where the simplest and cheapest is cold uniaxial pressing); and (3) solid-state sintering (i.e. heat treatment at high temperature, below the melting point of Ti) generally performed under vacuum [46]. This technique is cost effective, allows for better control of powder size and introduction of desirable pores. When it comes to pore size and shape, these are related to the size of the starting powder, its shape, its size and the shape of the spacer used to promote porosity. Among all of them, Mg is a good candidate in the manufacture of Ti scaffolds, as its solubility in Ti is low. Furthermore, Mg^{2+} increases osteoconductivity [47] and does not present any type of biomedical inconvenience, such as toxicity.

The main works carried out using Mg are for Ti-Al-V alloys [48]. Wen et al., (2001) reported that Mg foam prepared with a porosity of 50% showed a compressive strength of 2.33 MPa, with Young's modulus of 0.35 MPa, respectively [49]. Zhuang et al. [50] also evaluated the mechanical properties of porous Mg manufactured by powder metallurgy and the scaffolds with porosity from 36 to 55% showed a Young's modulus value in the range of 3.6 to 18.1 GPa, closer to that of natural bone [50]. They also investigated the effect of porosity on biodegradation. In their study, it was reported that materials with greater porosity degraded more quickly, due to greater interconnectivity and surface exposure, conditions that maximize chemical reactions. Although Mg materials have a low elastic modulus, mechanical resistance and corrosion are limiting factors for their use [51]. However, the use of Mg together with other elements to form porous alloys such as Ti can be an interesting alternative. For the production of porosity using this technique, it is necessary to control parameters such as temperature and times of the sintering steps, in addition to the particle size of the powders, due to its commitment to mechanical properties [52]. Such variables strongly influence the morphology of the pores, that is, it can provide the same amount of porosity, but with different shapes and sizes. The mechanical properties are also affected, mainly those related to the ductility and dynamic properties of the material, as they depend on the porosity characteristics [53]. The pores attenuate the applied force, do not distribute it over a larger area and cause local stress accumulation, so that they can even serve as sites for crack nucleation [54]. The effect of porosity on mechanical properties depends mainly on the following factors: volume fraction of the pores and their interconnection, size, morphology and distribution. The most important parameters are the total porosity, the shape of the pores and/or contacts during sintering [55].

Porosity has a noticeable and well-recognized effect on mechanical properties. Porosity can increase stress concentration and cause fractures. It was demonstrated by Danninger et al. that this parameter is directly related to the mechanical properties of an alloy [56]. These factors can be controlled by adjusting the sintering parameters, compaction pressure and particle size [57]. Optionally, functional porosity can be introduced by adapting the particle size of the starting powders and the sintering conditions. In addition, powder metallurgy allows flexibility in alloy design, mixing pure Mg powders with different elemental or alloy powders. Due to the high affinity of Mg for oxygen, all handling of powders and samples, as well as subsequent sintering, must be carried out under a protective atmosphere of argon or under vacuum [58] conditions, residual pores can vary between 2% and 45%. When approaching porosities close to 45%, interconnection (percolation) of the pores appears.

3.2 Fabrication via additive manufacturing

In recent years, interest has increased in the application of additive manufacturing of Mg alloys for its biomedical application. This can allow the obtaining of complex shapes adapted to the patient, since it would be a personalized manufacture. However, and due to the physical properties of Mg and its alloys, the application of additive manufacturing by melting the alloy has not been easy, since the boiling temperature of magnesium is very low (~ 1091°C). Despite this, and due to the manifest interest in the biodegradation properties of Mg motivate its application as biodegradable implants [59], seeking the best combination between resistance to corrosion, wear, mechanical properties and biocompatibility [60].

Another factor influencing the scarce development of additive techniques in Mg alloys is the ease of obtaining coupling parts by injection casting processes or the extrusion capacity of these alloys despite the difficulty imposed by their hcp crystalline structure. In addition, the great reactivity with oxygen that Mg presents must be considered and also limits the application of rapid heating techniques that could cause the combustion of the metal. However, some of the technologies applied to other materials have been applied, by means of specialized teams that require work in protective atmospheres, which ensure the possibility of handling these alloys [61].

The potentially most interesting techniques for the manufacture of magnesium alloys are powder bed fusion (PBF), especially selective laser melting (SLM), widely used in the development of different Mg alloys [62, 63]. Powder bed fusion (PBF) is an AM process in which thermal energy is used to selectively fuse regions of a powder bed [64]. The powder bed contains metal, polymer, or ceramic powder as feedstock. An energy source directed towards the powder bed selectively scans and melts the top layer of the powder bed. The powder bed then lowers and a fresh layer of powder is spread over the melted layer. This process continues until the entire structure has been formed by stacking melted layers of powder.

Another way to use additive manufacturing related to magnesium alloys is to have a porous structure obtained by additive manufacturing as in the case of Perets et al. [65] who obtain the TI-6Al-4 V mesh by SLM and then infiltrate the Mg elemental into the holes. In this way they can obtain structures that do not collapse, although their resistance is not increased. The magnesium will present an accelerated corrosion by galvanic effect and depending on the size of the pores, an osseointegration will be available as it is a very biocompatible set.

Degradation capability of Mg gives a feature of bioactivity in bone formation that leading Balog et al. develop a bioactive metal system compound by structural material for dental implants, via extrusion from a powder mixture of Ti and Mg (4 and 12%) in low temperature [66]. Adding Mg, is possible to obtain a bioactive system and a decreasing of elastic modulus, further promote a good osteointegration due to the Mg resorption and the presence of pores where the bone ingrowth can be formed.

Mechanical properties in Ti parts that receive infiltration of Mg depends on the amount of Mg and matrix used. Studies published by Jiang et al. about infiltration of Mg in a scaffold of Ti-Mg (99,9%) was possible to control density from compaction pressure with volume fractions up to 60% Ti, which confers stiffness similar to those of cortical bone [67]. The use of non-degradable Ti matrices, as described in previous sections, is necessary as a non-degradable support due to its excellent biocompatibility, high resistance to corrosion and excellent mechanical properties. Similarly, efforts have been made to obtain porous Ti by means of spacer techniques [68] or by additive manufacturing processes such as selective laser melting (SLM) [69, 70]. Biodegradable Mg-based alloys are advantageous as fillers for bioactive implants because the release of Mg ions during corrosion *in vivo* is non-toxic and, furthermore, may have a beneficial effect on tissue regeneration and osteoblast response [71]. However, excessive in vivo corrosion of Mg can result in a premature loss of mechanical integrity and it is for this reason that a balance is struck between the structural integrity provided by titanium alloys and the degradation of Mg alloys. The aim is therefore to achieve synergy between both alloys and therefore to improve or control the biodegradation of magnesium through alloys that allow its corrosion rate to be controlled and can adapt it to the rate of bone growth, as proposed by Perets et al. with Mg-2.4% Nd-0.6% Y-0.3% Zr [65] alloy infiltration working in simulated environments.

4. Microstructural features of Ti-Mg alloys

The Ti-alloys are already recognized as the most promising materials for dental and orthopedic implants. This is due to their excellent biocompatibility, good mechanical properties, superior corrosion resistance, and no allergic issues [72]. Adding Mg provides multiple biological advantages to Ti biomaterials. Some of these advantages are an elastic modulus closer to that of human bone, high biocompatibility, and low toxicity. The mismatch between elastic modules of human bone and dense metallic biomaterials is considered the main cause of implant failure. Typically, the elastic modulus of human bone has a maximum value of 30 GPa, while that for metallic biomaterials ranges from 100 GPa for chemically pure Ti to

230 GPa for Co-Cr alloy [15]. From the metallic biomaterials, the near β-Ti alloys have reported some of the lowest elastic modulus values ranging from 80 to 110 GPa [14]. Additionally, porous β-Ti alloys have obtained elastic modulus ranging from 43 to 75 GPa [4, 37, 73, 74], and it is expected that the addition of Mg will decrease even more such values [75]. An elastic modulus near to that of human bone improves the performance of multiple biomaterials used for dental and orthopedic implants. Considering that the microstructural effects of Mg additions are important to define the mechanical behavior of biomaterials, it is important to study the microstructure.

Due to the above, the effect of low (3 mass%) Mg additions into a Ti-34Nb-6Sn alloy was studied [28]. The selection of Nb and Sn was done for its β-stabilizing effect on Ti alloys [9, 76]. The β-phase has shown better mechanical compatibility with the human bone in comparison to the α-phase. This is due to lower elastic modulus values and a high strength-to-weight ratio [13, 14]. An elastic modulus of the implant that is close to that of the human bone decreases the mismatch of mechanical stress through the interface and avoids the damage of the organic tissue cells. Based on the above, the closer the elastic modulus between both, bone and implant material, the lower the probability of crack nucleation and failure of the implant [72]. Furthermore, a high strength-to-weight ratio allows reducing the thickness of biomedical metallic implants. Besides, the selected contents of Nb and Sn also contribute to obtaining elastic modulus of the Ti alloy of ~60 GPa [9, 74, 77]. The addition of 6 wt.% of Sn into the Ti-Nb alloy showed a good combination of corrosion resistance, strength, hardness, and lower elastic modulus [4]. On the other hand, considering the low elastic modulus of Mg (from 39 to 46 GPa) [78], it was added to reduce the elastic modulus of the Ti alloy. Besides, Mg is a natural component of human bones and is a required element for the metabolism process, i.e., Mg exhibits great biocompatibility, non-toxicity, and can stimulate hard tissue recovery [2]. The reported biodegradability of Mg^{2+} is one more of its valuable advantages [79]. The possibilities of its use in dental and orthopedic implants can be highly beneficial from controlling its degradation rate and ensuring its mechanical integrity during desire clinical periods. Moreover, the corrosion products of Mg, trigger the osteoconductivity of the bone [75]. For the last, the β-phase Ti alloys have shown a superior electrochemical performance that provides better resistance in corrosive environments as oral or body fluids [13, 32]. This is due to the surface TiO_2 passive film that inherently protects these alloys [10, 32].

From the above-mentioned selection of components, four Ti alloys were prepared by powder metallurgy method. A typical four-stage route of milling – mixing - compaction – sintering was used. For this, measured amounts of titanium hydride (TiH), niobium hydride (NbH), and atomized Sn were used to obtain a mass ratio of Ti, Nb, and Sn of 60:34:6. The powders were mixed in a planetary ball mill and grounded at 200 rpm for 40 min. Posteriorly, the mixed powders were dried under a vacuum. Details of processing parameters can be found in previous work [28]. Half of the mixed powder was saved with the abovementioned chemical composition, while the other half was mixed with Mg powder in a 3 mass%. All the dried powders, whether with or without Mg, were compacted at 200 MPa for 15 s. To evaluate the microstructural and mechanical effect of Mg addition, two different sintering temperatures were used, 900°C and 1100°C. The sintering was carried out in a high vacuum resistive furnace for 2 h. A scheme of the elaboration process is shown in **Figure 1**.

Finally, two Ti-34Nb-6Sn (TNS) and two Ti-34Nb-6Sn/Mg (TNS/M) alloys were produced. The identification of the obtained alloys indicates the sintering temperature as postfix: TNS_{900}, TNS_{1100}, TNS/M_{900}, and TNS/M_{1100}.

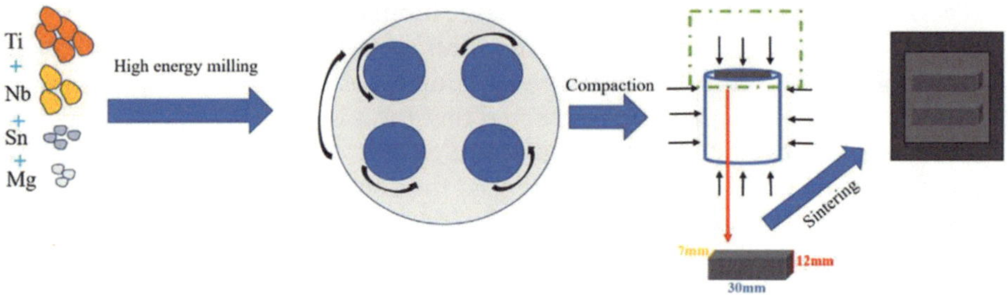

Figure 1.
Representation of the methodology to elaborate the Ti-34Nb-6Sn and Ti-34Nb-6Sn/Mg alloys [28].

For microstructural analysis, the samples were subjected to conventional metallographic preparation until a mirror-like surface. Final polishing with oxide polishing suspension (OPS) solution and hydrogen peroxide (10:2) was applied. Rietveld refinement of X-ray diffraction measurements was carried out for quantifying the present phases and estimating the lattice parameters. A Bruker/D2Phaser with Cu-Kα radiation was used at 30 kV and 10 mA. The measured 2θ range was 20 and 90° with a step size of 0.02° every 10 s. The Rietveld refinement was carried out by MAUD software (version 2.94) [80]. The morphology and chemical distribution of phases were studied by field emission scanning electron microscopy (FESEM) (ZEISS-ULTRA 55) and an energy-dispersive X-ray spectroscopy detector (EDS) (Oxford Instruments Ltda.).

Considering that the mechanical properties play a determining role in the performance of biomaterials, the elastic modulus was estimated by impulse excitation technique (ATCP, Sonelastic®). Hardness measurements were obtained using a load of 147 N by the Rockwell method (BECLA), using a spherical steel indenter with a diameter of 0.16 cm. More details of the whole methodology can be found in previous works [28–31].

As result, the four alloys resulted in tri-phasic microstructures of α-Ti (under hexagonal compact (hcp), structure), β-Ti (under body centered cubic (bcc) structure), and segregation of Nb. The microstructures can be observed in **Figure 2**, where the light gray color corresponds to the matrix of β-phase, the dark gray to the α-phase, and the bright particles to the Nb segregation. As it was expected from the increment of temperature, bigger grain sizes can be observed in the samples sintered at 1100°C (**Figure 2c** and **d**) compared to those sintered at 900°C. Both, α and β-phases, are randomly distributed in the microstructure, however, the linear chemical composition through the microstructure is not homogeneous. This is due to lower contents of Nb and Sn, especially Nb, in the α-phase (**Figure 2e**). Due to the well-known β-stabilizer nature of both alloying elements [1, 3–10], those chemical gradients were expected. On the other hand, the Nb segregation occurrence was reduced with the sintering temperature (**Figure 2c** and **d**), which indicates a better Nb diffusion in the matrix when temperature increases. However, the Nb particles are continuously observed at both sintered temperatures. From the Ti-Nb diagram phase, Nb has low solubility in Ti, so the continuous presence of Nb segregates was expected [3].

The phases percentages estimated by Rietveld refinement from XRD measurements and the total porosity obtained by Archimedes method for the four alloys are presented in **Table 1**. The TNS_{900}, TNS_{1100}, and TNS/M_{1100} samples showed similar phases percentages. However, the TNS/M_{900} showed a reduced β to α transformation during sintering. Additionally, both TNS/M samples obtained higher porosity

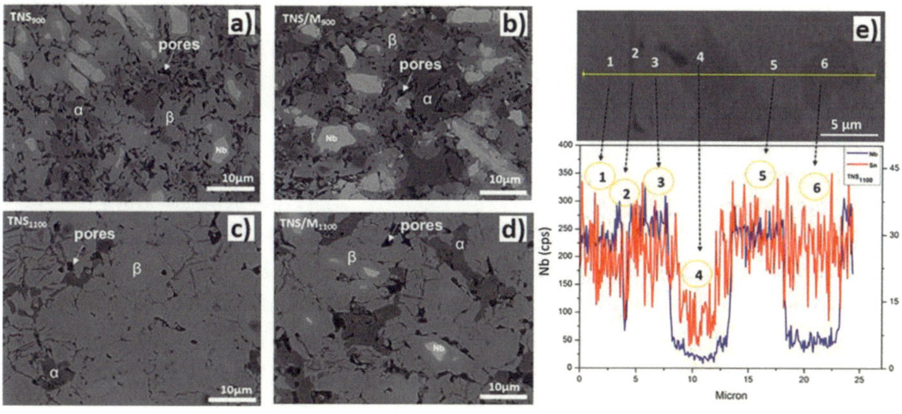

Figure 2.
Microstructure of a) TNS900, b) TNS/M900, c) TNS1100, and d) TNS/M1100, as well as e) linear chemical gradients through α and β phases representative of the four studied samples. Adapted from [28].

Sample	α (mass%)	β (mass%)	Total porosity (%)
TNS_{900}	20.1 ± 0.3	79.9 ± 0.0	22 ± 1
TNS_{1100}	23.8 ± 0.5	76.1 ± 0.0	11 ± 1
TNS/M_{900}	35.3 ± 0.0	64.7 ± 0.0	29 ± 1
TNS/M_{1100}	22.4 ± 0.0	77.6 ± 1.4	20 ± 0.5

Table 1.
Phases percentages and total porosity of TNS and TNS/M samples sintered at 900°C and 1100°C.

percentages in comparison with the Mg-free alloys. This was a clear suggestion about the reduction of diffusional processes when Mg is added into the Ti-Nb-Sn system. Thus, Mg addition has an apparent α-phase stabilization effect.

Furthermore, the increment of porosity with the Mg content could be related to the highest content of oxygen from the intrinsic passivation layer of Mg^{2+}. When temperatures increase, the release of gas also increases, generating pores at the microstructure [81]. Additionally, the Mg powders acted as a spacer in the TNS/M samples. Comparing the low melting point of Mg (~650°C) with that of Ti (~1668°C), it is evident that a fraction of Mg is evaporated during sintering, while the Ti content remains constant. The partial evaporation of Mg assisted in the formation of the pores during sintering. It is well known that the porosity tends to decrease at higher sintering temperatures during powder metallurgy methods [82]. Thus, the reduced porosity for the samples sintered at 1100°C compared to the samples sintered at 900°C was an expected result. The porosity could influence the mechanical properties of the alloy, especially in the strength and elastic modulus. The mechanical properties will be discussed in the next Section 3.

Posteriorly, the previous Ti-34Nb-6Sn and Ti-34Nb-6Sn/Mg alloys were also reported through the same powder metallurgy methodology, except for sintering temperatures of 700°C and 800°C [29–31]. These samples will be identified as TNS_{700}, TNS/M_{700}, TNS_{800}, and TNS/M_{800} for this book chapter. Compared to the previous TNS/M_{900}, and TNS/M_{1100}, the same three constitutive phases, α, β and Nb segregation, were observed at the TNS and TNS/M alloys sintered at 700°C and 800°C. For comparison purposes, representative EDS measurements of TNS/M_{800} and TNS/M_{900} are shown in **Figure 3**. **Figure 3a** is representative of the distribution of the elements in the samples sintered at 700°C and 800°C, while **Figure 3b** represents the distribution of the elements in samples sintered at 900°C and 1100°C. Similar distributions of the alloying elements were observed in both cases, except for the Nb segregates. **Figure 3a** shows a greater presence of Nb segregates in comparison with that representative of sintering above 900°C (**Figure 3b**). This can be explained by the lower solubility of Nb in Ti below 1100 K (~827°C) [3]. The solubility, together with the effect of temperature, could also be related to the smaller grain size in the samples sintered below 900°C (**Figures 2** and **3**). While the β-phase matrix of the TNS_{1100} has an average grain size of 15 μm, that of the TNS_{700} has an average of 6 μm. This could be due to the lower solubility of Nb at lower sintering temperatures generated more segregated particles through the microstructure. Those particles could act as a pin for grain growth. The pin-like behavior of Nb segregates in a Ti matrix has been reported before [16].

From **Table 2**, the TNS_{700}, TNS/M_{700}, TNS_{800}, and TNS/M_{800} samples presented higher porosity percentages in comparison with the samples sintered at 900°C and 1100°C (**Table 1**). Besides, the samples sintered at 800°C were more compacted than the samples sintered at 700°C. This was a confirmation about the higher the sintering temperature, the lower the porosity percentage. It is well-known that increasing the sintering temperature favors the diffusional processes and more compacted microstructures with higher relative densities, i.e., lower porosity percentages [82]. Besides, **Table 1** and **Table 2** showed an increment of porosity with the Mg additions for samples sintered at the same temperature. As it was explained before, the partial evaporation of Mg and its passivating oxide, contributed to the increment of porosity. Among the most studied spacers for powder metallurgy methods are carbamide, sodium chloride, ammonium hydrogen carbonate, and Mg [83]. From those, Mg has shown superior advantages over the organic spacers due to its good biocompatibility

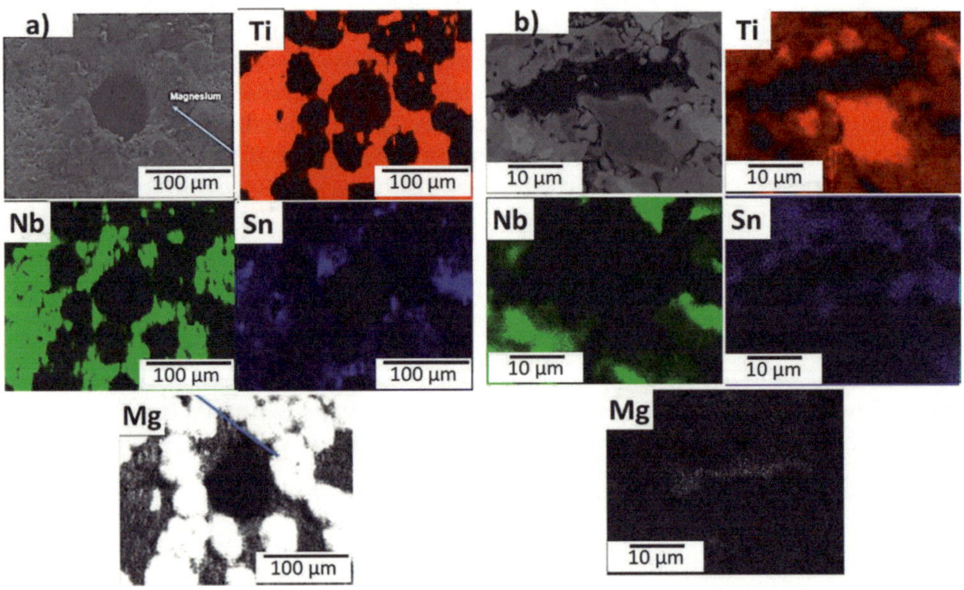

Figure 3.
Comparison between alloying elements distribution of the Ti-34Nb-6Sn/Mg alloy representative of a) the sintered at 700°C and 800°C and b) the sintered at 900°C and 1100°C. adapted from [28, 30].

Sample	α (mass%)	β (mass%)	Total porosity (%)
TNS_{700}	45 ± 0.3	55 ± 0.3	23 ± 0.5
TNS_{800}	31 ± 0.2	69 ± 0.2	21 ± 1.4
TNS/M_{700}	41 ± 0.1	59 ± 0.1	38 ± 0.6
TNS/M_{800}	33 ± 0.3	67 ± 0.3	28 ± 0.3

Table 2.
Phases percentages and total porosity of TNS and TNS/M samples sintered at 700 °C and 800 °C.

and good mechanical properties [79]. The increment of pore formation with the Mg additions could be beneficial for the implant by decreasing the elastic modulus. This topic will be covered in Section 3.

Besides, from **Figure 3**, a larger pore size for the samples sintered below 800°C is notable in comparison with that for the samples sintered above 900°C. Macropores of ~100 μm in average were acquired (**Figure 3a**) for the samples sintered below 800°C. This was contrasting with the pores in the range from 5 to 35 μm obtained in the samples sintered above 900°C. This could be related to lower atomic diffusion resulting from the lower sintering temperature that reduced the bonding ratio in the samples. Porosity improves the bonding between bone and implant material, encouraging the anchorage and growth of the organic tissue [15, 75]. Thus, porous materials facilitate tissue generation enabling body fluid transmission [84]. Considering that allowing successful osseointegration is one of the main requirements of dental and orthopedic implant materials, the porous structures are highly promising for those applications. It has also been reported that macro-pores are beneficial for multiple biological processes as cell attachment, ingrowth of osteoblasts, vascularization, and osteoconductivity [85]. However, an adequate vascularization requires pores with diameters larger than 100 μm, specifically in the range from 100 to 500 μm [15, 86]. From the above and the fact that increments of porosity percentage used to assist on the decrement of elastic modulus [75], the alloys sintered at temperatures below 800°C could be more appropriate for dental or orthopedic implant applications. However, other mechanical properties as strength and hardness are also crucial for the performance of metallic implants. The mechanical behavior will be described in the following Section 5. This will contribute to clarifying the current concerns about the role of porosity *in-vivo* environments.

5. Mechanical behavior of Ti-Mg alloys

It is well known that the microstructural features have a strong effect on the mechanical behavior of metallic components. In this section, the mechanical performance of the Ti-34Nb-6Sn and Ti-34Nb-6Sn/Mg alloys will be explained based on their microstructural features described in Section 4.

As it was explained before, the Mg additions into dental or orthopedic biomaterials should assist the adequate biocompatibility and a good mechanical strength-elastic modulus relationship. Lower elastic modulus is expected with Mg additions. Moreover, lower elastic moduli are expected from higher porosity percentages, this is, for lower sintering temperatures. Simultaneously, higher hardness values can be expected from the more compacted samples, i.e., the less porous. To evaluate these hypotheses, hardness and elastic modulus measurements were carried out.

From previous works [28–30], **Table 3** presents the Vickers hardness as a function of sintering temperature for the TNS and TNS/M samples. A general tendency to increase hardness as a function of sintering temperature was observed. This is congruent with the lower total porosity percentage in the samples sintered at higher temperatures. As it was explained in Section 2, the temperature increment fabrication atomic diffusional processes that result in bonding improvement and density increment. To observe the effect of temperature on the total porosity and hardness of the sintered alloys, **Figure 4** compares those three values. A general decrement of porosity with the increment of temperature can be observed in the TNS (**Figure 4a**) and TNS/M (**Figure 4b**) samples. Simultaneously, the higher density of the samples resulted in higher hardness values for all the samples.

For comparison purposes, the hardness of elemental Ti ranges from 1.3 to 2.0 GPa [16], while the hardness in the TNS and TNS/M samples ranges from 0.9 to 4.0 GPa. Besides, the minimum recommended hardness value for metallic biomaterials is 1.2 GPa for avoiding high wear damage susceptibility during chewing and daily oral processes [87]. However, the hardness of the natural human teeth ranges from 2.2 to 3.9 GPa [88]. Hardness values near to the ones of natural teeth could assist in decreasing wearing between teeth and implant. From the sintered TNS and TNS/M alloys, the TNS_{700}, TNS_{800}, TNS/M_{700}, and TNS/M_{800} have hardness values below 1.5 GPa, this is, below the minimum acceptable for avoiding wear damage and being within the range of the hardness of natural teeth. As result, those four alloys cannot serve as a feasible biomaterial for dental applications. However, the hardness of the human bone ranges from ~0.3 to ~0.75 GPa [89, 90]. This means that all the sintered samples are within the acceptable hardness range for being applied as orthopedic implants.

From **Figure 4b**, it is also possible to observe a slower decrement rate of total porosity with the sintering temperature in comparison with that for the samples free of Mg additions (**Figure 4a**). This could be related to the abovementioned effect of Mg as a spacer. The partial evaporation of Mg created additional pores compared to the created in the Mg-free samples (TNS). As result, the hardness increment with sintering temperature in Mg-added (TNS/M) alloys also showed a slower rate of increment in comparison with the TNS samples.

As it was discussed in previous Section 4, the elastic modulus plays a key role in the success rate of dental and orthopedic implants. Considering that the elastic modulus is a measure of the stiffness of the material, it determines the resistance to

Sample	Hardness (HV)
TNS_{700}	146 ± 16
TNS_{800}	153 ± 12
TNS_{900}	309 ± 44
TNS_{1100}	411 ± 25
TNS/M_{700}	92 ± 10
TNS/M_{800}	120 ± 20
TNS/M_{900}	226 ± 85
TNS/M_{1100}	344 ± 24

Table 3.
Comparison of Vickers hardness (HV) between the TNS and TNS/M samples sintered at 700 °C, 800 °C, 900 °C, and 1100 °C.

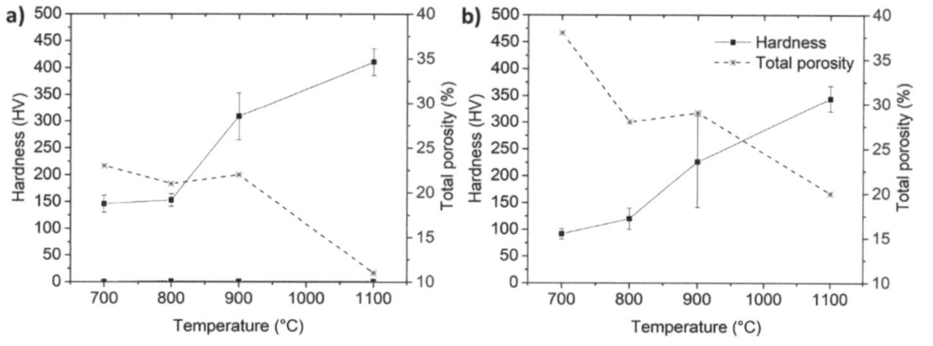

Figure 4.
Hardness and total porosity as a function of sintering temperature for the a) TNS and b) TNS/M samples.

deform a material in the elastic range. For increasing the feasibility of the implant, the constituent alloy should have an elastic modulus near to that of the human bone. The elastic modulus of human bone ranges from 5 to 30 GPa [15, 91].

For evaluating the elastic modulus of the sintered alloys, **Figure 5** presents a comparison between the values obtained for the TNS and TNS/M samples sintered at different temperatures. Considering that the samples sintered at 700°C and 800°C were discarded as potential materials for dental implant applications, the samples TNS_{700}, TNS_{800}, TNS/M_{700}, and TNS/M_{800} were not included in **Figure 5**. Lower elastic modulus can be observed in the samples with Mg addition compared to the TNS systems. This result was congruent with the lower hardness values of the TNS/M samples (**Figure 4**), which implies that these alloys are softer than the TNS for similar sintering conditions. As it was explained before, the lower hardness in the TNS/M samples resulted from the spacer-like behavior of the Mg powders. Besides, a tendency to increase the elastic modulus with the sintering temperature was observed. Being congruent with lower porosity percentages measured in the TNS_{1100} and TNS/M_{1100} samples.

Comparing the obtained elastic moduli in the studied samples with the ones reported for human bone, the TNS/M_{900} sample can be the most adequate for its use in dental or orthopedic implants. Besides, the TNS/M_{900} alloy joins an acceptable hardness for biomedical implants and has adequate porosity features for triggering the anchorage between organic tissue and implant material. This is, the TNS/M_{900} sample

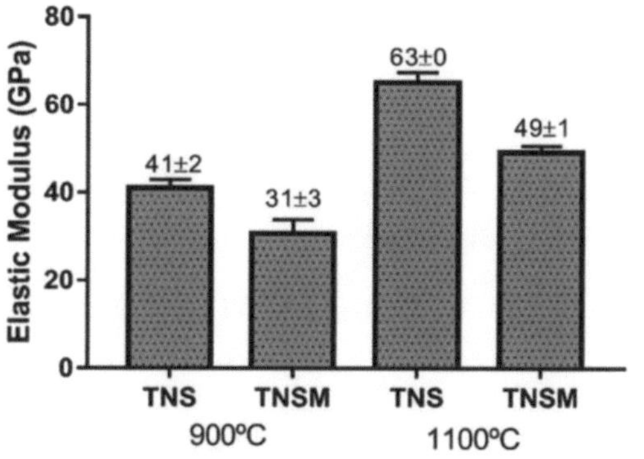

Figure 5.
Elastic modulus as function of temperature for samples sintered at 900°C and 1100°C with (TNSM) and without (TNS) Mg addition [28].

combined the best microstructural and mechanical properties to be a potential biomaterial. However, the performance of these alloys under in vivo environments should be described to determine the feasibility of the studied alloys for biomedical purposes.

6. Conclusion and future perspectives

Development of biomaterials needs to focus on the biointerface construction to match the structure of the host tissue and to meet the mechanical requirements of specific tissue. In order to do that, metallurgic and additive manufacturing techniques present great potential in the development of Ti-Mg alloys, with complex shape formed by pores to be more biocompatible. It is critical to manipulate the surface by physical and chemical parameters to achieve the clinical purpose of the biomaterial, leading to a fast integration to the bone, due to the stimulating biological functions. In this review was showed that β -Ti-Nb-Sn alloy can be fabricated using Mg to create high content of porosity (20–38%), with an elastic modulus between 31 and 49 GPa close to bone tissue, and hardness close to commercial materials and higher than different parts of skeleton.

In this sense continuous studies and researches in this field is of great relevance for materials applied in life sciences.

Acknowledgements

This work was supported by the São Paulo State Research Support Foundation (FAPESP) [grants: 2017/13876-2; 2019/24237-6], by the Ministerio Español de Ciencia, Innovación y Universidades with Grant RTI2018-097810-B-I00 and the São Paulo State Institute for Technological Research, in the development of materials, whom the authors thank.

Conflict of interest

The authors declare no conflict of interest.

Author details

Mariana Correa Rossi[1*], Liliana Romero Resendiz[2] and Vicente Amigó Borrás[1]

1 Polytechnic University of Valencia, Valencia, Spain

2 Materials Research Institute – UNAM, Mexico

*Address all correspondence to: mrncorrea90@gmail.com

References

[1] Westerman RW, Scammell BE. Principles of bone and joint injuries and their healing. Surgery. 2012;**30**:54-60. DOI: 10.1155/2018/9216314

[2] Tan L, Yu X, Wan P, Yang K. Biodegradable materials for bone repairs: A review. Journal of Materials Science and Technology. 2013;**2013**(29):503-513. DOI: 10.3390/ma8095273

[3] Bram, Dr M, Ebel Dr T, Wolff M, et al. Applications of powder metallurgy biomaterials. In: Chang I, Zha Y, editors. Advances in Powder Metallurgy. 1st ed. Woodhead publishing; 2013. pp. 520-554. DOI: 10.1533/9780857098900.4.520

[4] Sezer N, Evis Z, Kayhan SM, et al. Review of magnesium-based biomaterials and their applications. J Magnes Alloy. 2018;**6**:23-43. DOI: 10.1016/j.jma.2018.02.003

[5] Chinnappan R, Panigrahi BK, van de Walle A. First-principles study of phase equilibrium in Ti–V, Ti–Nb, and Ti–Ta alloys. Calphad Comput Coupling Phase Diagrams Thermochem. 2016;**54**:125-133. DOI: 10.1016/j.calphad.2016.07.001

[6] Moraes PEL, Contieri RJ, Lopes ESN, et al. Effects of Sn addition on the microstructure, mechanical properties and corrosion behavior of Ti-Nb-Sn alloys. Materials Characterization. 2014;**96**:273-281. DOI: 10.1016/j.matchar.2014.08.014

[7] Mohan P, Rajak DK, Pruncu CI, et al. Influence of β-phase stability in elemental blended Ti-Mo and Ti-Mo-Zr alloys. Micron. 2021;**142**:102992. DOI: 10.1016/j.micron.2020.102992

[8] Elshalakany AB, Ali S, Amigó Mata A, et al. Microstructure and Mechanical Properties of Ti-Mo-Zr-Cr Biomedical Alloys by Powder Metallurgy. Journal of Materials Engineering and Performance. 2017;**26**:1262-1271. DOI: 10.1007/s11665-017-2531-z

[9] Mohan P, Elshalakany AB, Osman TA, et al. Effect of Fe content, sintering temperature and powder processing on the microstructure, fracture and mechanical behaviours of Ti-Mo-Zr-Fe alloys. Journal of Alloys and Compounds. 2017;**729**:1215-1225. DOI: 10.1016/j.jallcom.2017.09.255

[10] Rossi MC, de Santi GE, Rodríguez MVH, et al. Study of the current density of the electrical resistance sintering technique on microstructural and mechanical properties in a β Ti-Nb-Sn ternary alloy. Applied Physics A: Materials Science & Processing. 2021;**127**:1-19. DOI: 10.1007/s00339-021-04937-4

[11] Zhang DC, Yang S, Wei M, et al. Effect of Sn addition on the microstructure and superelasticity in Ti-Nb-Mo-Sn Alloys. Journal of the Mechanical Behavior of Biomedical Materials. 2012;**13**:156-165. DOI: 10.1016/j.jmbbm.2012.04.017

[12] Atapour M, Pilchak AL, Frankel GS, et al. Corrosion behavior of β titanium alloys for biomedical applications. Materials Science and Engineering: C. 2011;**31**:885-891. DOI: 10.1016/j.msec.2011.02.005

[13] Llobet JM, Domingo JL. Acute toxicity of vanadium compounds in rats and mice. Toxicology Letters. 1984;**23**:227-231. DOI: 10.1016/0378-4274(84)90131-0

[14] Costa BC, Tokuhara CK, Rocha LA, et al. Vanadium ionic species from degradation of Ti-6Al-4V metallic implants: In vitro cytotoxicity and speciation evaluation. Materials Science and Engineering: C. 2019;**96**:730-739

[15] Bania PJ. Beta titanium alloys and their role in the titanium industry. JOM. 1994;**46**:16-19. DOI: 10.1007/BF03220742

[16] Veiga C, Davim JP, Loureiro A. Properties and applications of titanium alloys: A brief review. Reviews on Advanced Materials Science. 2012;**32**:14-34

[17] Goia TS, Violin KB, Yoshimoto M, et al. Osseointegration of titanium alloy macroporous implants obtained by PM with addition of gelatin. Adv Sci Technol. 2010;**76**:259-263. DOI: 10.4028/www.scientific.net/AST.76.259

[18] Romero-Resendiz L, Gómez-Sáez P, Vicente-Escuder A, et al. Development of Ti-In alloys by powder metallurgy for application as dental biomaterial. Journal of Materials Research and Technology. 2021;**11**:1719-1729. DOI: 10.1016/j.jmrt.2021.02.014

[19] Lario J, Vicente Á, Amigó V. Evolution of the microstructure and mechanical properties of a Ti35Nb2Sn alloy post-processed by hot isostatic pressing for biomedical applications. Metals (Basel). 2021;**11**:1-8. DOI: 10.3390/met11071027

[20] Amigó-Mata A, Haro-Rodriguez M, Vicente-Escuder Á, et al. Development of Ti–Zr alloys by powder metallurgy for biomedical applications. Powder Metallurgy. 2021;**65**:1-8. DOI: 10.1080/00325899.2021.1943182

[21] Zambrano Carrullo JC, Dalmau Borrás A, Amigó Borrás V, et al. Electrochemical corrosion behavior and mechanical properties of Ti–Ag biomedical alloys obtained by two powder metallurgy processing routes. Journal of the Mechanical Behavior of Biomedical Materials. 2020;**112**:104063. DOI: 10.1016/j.jmbbm.2020.104063

[22] Mostaed E, Hashempour M, Fabrizi A, et al. Microstructure, texture evolution, mechanical properties and corrosion behavior of ECAP processed ZK60 magnesium alloy for biodegradable applications. Journal of the Mechanical Behavior of Biomedical Materials. 2014;**37**:307-322. DOI: 10.1016/j.jmbbm.2014.05.024

[23] Song G. Control of biodegradation of biocompatable magnesium alloys. Corrosion Science. 2007;**49**:1696-1701. DOI: 10.1016/j.corsci.2007.01.001

[24] Walker J, Shadanbaz S, Woodfield TBF, et al. Magnesium biomaterials for orthopedic application: A review from a biological perspective. J Biomed Mater Res - Part B Appl Biomater. 2014;**102**:1316-1331. DOI: 10.1002/jbm.b.33113

[25] Wang J, Ma XY, Feng YF, et al. Magnesium Ions Promote the Biological Behaviour of Rat Calvarial Osteoblasts by Activating the PI3K/Akt Signalling Pathway. Biological Trace Element Research. 2017;**179**:284-293. DOI: 10.1007/s12011-017-0948-8

[26] Nie X, Sun X, Wang C, et al. Effect of magnesium ions/Type i collagen promote the biological behavior of osteoblasts and its mechanism. Regen Biomater. 2019;**7**:53-61. DOI: 10.1093/rb/rbz033

[27] Wang G, Ge S, Shen Y, et al. Study on the biodegradability and biocompatibility of WE magnesium alloys. Materials Science and Engineering: C. 2012;**32**:2190-2198

[28] Wang W, Wu H, Zan R, et al. Microstructure controls the corrosion behavior of a lean biodegradable Mg–2Zn alloy. Acta Biomaterialia. 2020;**107**:349-361. DOI: 10.1016/j.actbio.2020.02.040

[29] Kirkland NT. Magnesium biomaterials: Past, present and future. Corrosion Engineering, Science and Technology. 2012;**47**:322-328. DOI: 10.1179/1743278212y.0000000034

[30] Rossi MC, Bayerlein DL, Gouvêa EDS, et al. Evaluation of the influence of low Mg content on the mechanical and microstructural properties of β titanium alloy. Journal of Materials Research and Technology. 2021;**10**:916-925. DOI: 10.1016/j.jmrt.2020.12.103

[31] Rossi MC. Caracterização da liga ti-sn-nb com mg e seu potencial biotecnológico na consolidação óssea [thesis]. São Paulo , Brazil: São Paulo State University; 2021

[32] da Silva DC, Rossi MC, EVP A, et al. Low Mg content on Ti-Nb-Sn alloy when in contact with eBMMSCs promotes improvement of its biological functions. Journal of Materials Science: Materials in Medicine. 2021;**32**:144. DOI: 10.1007/s10856-021-06620-9

[33] Rossi MC, Bayerlein DL, de Souza Brandãoa J, et al. Physical and biological characterizations of TiNbSn/(Mg) system produced by powder metallurgy for use as prostheses material. Journal of the Mechanical Behavior of Biomedical Materials. 2021;**115**:104260. DOI: 10.1016/j.jmbbm.2020.104260

[34] Liu C, Ren Z, Xu Y, Pang S, Zhao X, Zhao Y. Biodegradable magnesium alloys developed as bone repair materials: A review. Scanning. 2018;**2018**:1-15. DOI: 10.1155/2018/9216314

[35] Witte F, Kaese V, Haferkamp H, Switzer E, Meyer-Lindenberg A, Wirth CJ, et al. In vivo corrosion of four magnesium alloys and the associated bone response. Biomaterials. 2005;**26**:3557-3563. DOI: 10.1016/j.biomaterials.2004.09.049

[36] Leidi M, Dellera F, Mariotti M, Banfi G, Crapanzano C, Albisetti W, et al. Nitric oxide mediates low magnesium inhibition of osteoblast-like cell proliferation. The Journal of Nutritional Biochemistry. 2012;**23**:1224-1229. DOI: 10.1016/j.jnutbio.2011.06.016

[37] Chaya A, Yoshizawa S, Verdelis K, Myers N, Costello BJ, Chou DT, et al. In vivo study of magnesium plate and screw degradation and bone fracture healing. Acta Biomaterialia. 2015;**18**:262-269. DOI: 10.1016/j.actbio.2015.02.010

[38] Pichler K, Kraus T, Martinelli E, Sadoghi P, Musumeci G, Uggowitzer PJ, et al. Cellular reactions to biodegradable magnesium alloys on human growth plate chondrocytes and osteoblasts. International Orthopaedics. 2014;**38**:881-889. DOI: 10.1007/s00264-013-2163-3

[39] Rosalbino F, De Negri S, Scavino G, Saccone A. Microstructure and in vitro degradation performance of Mg-Zn-Mn alloys for biomedical application. Journal of Biomedial Materials Research Part A. 2013;**101**:704-711. DOI: 10.1002/jbm.a.34368

[40] Staiger MP, Pietak AM, Huadmai J, Dias G. Magnesium and its alloys as orthopedic biomaterials: A review. Biomaterials. 2006;**27**:1728-1734

[41] Fawcett WJ, Haxby EJ, Male DA. Magnesium: Physiology and pharmacology. British Journal of Anaesthesia. 1999;**83**:302-320

[42] Clarke K, Kashiwaya Y, King MT, Gates D, Keon CA, Cross HR, et al. The β/α peak height ratio of ATP a measure of free [Mg^{2+}] using 31P NMR. The Journal of Biological Chemistry. 1996;**271**:21142-21150. DOI: 10.1074/jbc.271.35.21142

[43] Kretsinger RH. Magnesium in biological systems. In: Kretsinger RH, Uversky VN, Permyakov EA, editors. Encyclopedia of Metalloproteins. New York, NY: Springer; 2013. DOI: 10.1007/978-1-4614-1533-6_258

[44] Pollak MR, Mount DB. Hereditary disorders of the thick ascending limb and distal convoluted tubule. In: Pollack MR, Mount DB, editors. Molecular and Genetic Basis of Renal Disease. 1st ed. Philadelphia, PA, USA: Saunders Elsevier; 2008. pp. 229-249. DOI: 10.1016/B978-1-4160-0252-9.50019-7

[45] Monteiro CP, Laires MJ. Magnesium in health and disease. In: Kretsinger RH, Uversky VN, Permyakov EA, editors. Encyclopedia of metalloproteins. New York: Springer; 2013. pp. 1264-1269

[46] Bolzoni L, Raynova S, Yang F. An alternative method to manufacture Ti alloys from particulate materials. Powder Technology. 2021;**280**:341-348. DOI: 10.1016/j.powtec.2020.11.013

[47] Yoshizawa S, Brown A, Barchowsky A, Sfeir C. Magnesium ion stimulation of bone marrow stromal cells enhances osteogenic activity, stimulating the effect of magnesium alloy degradation. Acta Biomaterialia. 2014;**10**:2834-2842. DOI: 10.1016/j.actbio.2014.02.002

[48] Kalantari SM, Arabi H, Mirdamadi S, Mirsalehi SA. Biocompatibility and compressive properties of Ti-6Al-4V scaffolds having Mg element. Journal of the Mechanical Behavior of Biomedical Materials. 2015;**48**:183-191. DOI: 10.1016/j.jmbbm.2015.04.015

[49] Wen CE, Mabuchi M, Yamada Y, Shimojima K, Chino Y, Asahina T. Processing of Biocompatible Porous Ti and Mg. Scripta Materialia, DOI. 2001;**45**:1147-1153. DOI: 10.1016/S1359-6462(01)01132-0

[50] Zhuang H, Han Y, Feng A. Preparation, mechanical properties and in vitro biodegradation of porous magnesium scaffolds. Materials Science and Engineering C. 2008;**28**:1462-1466. DOI: 10.1016/j.msec.2008.04.001

[51] Wittea F, Kaese V, Haferkamp H, SwitzercE, Meyer-LindenbergA, WirthCJ, et al. In vivo corrosion of four magnesium alloys and the associated bone response. Biomaterials. 2005;**26**:3557-3563. DOI: 10.1016/j.biomaterials.2004.09.049

[52] Stoyanova V. Mechanical properties of high density low alloyed pm steels: Effect of sintering and secondary heat treatments [thesis]. Trento, Italy: University of Trento; 2005

[53] Stoyanova V et al. Influence of microstructure and porosity on static and dynamic mechanical properties of High Performance PM steels. In: Proceedings of the Euro PM. Vienna, Austria. 2004:47-62

[54] Molinari A, Menapace C, Santuliana E, Straffelini G. A simplified model for the impact resistance of porous sintered steels. Powder Metallurgy Progress. 2011;**11**:12-20

[55] Blanco L, Campos M, Torralba JM, Klint D. Quantitative Evaluation of Porosity Effects in Sintered and Heat Treated High Performance Steels. Powder Metallurgy, DOI. 2005;**48**:315-322. DOI: 10.1179/174329005X82199

[56] Danninger H, Spoljaric D, Weiss B. Microstructural Features Limiting the Performance of P/M Steels. International Journal of Powder Metallurgy. 1997;**53**: 43-53

[57] Maroli B et al. Sinter-Hardening and Heat Treatments of Materials Based on Astaloy CrM. Vol. 5. Las Vegas, USA: Proceedings of the PM2TEC 2003 International Conference on Powder Metallurgy & Particulate Materials; 2003. pp. 149-161

[58] Wolff M, Dahms M, Ebel T. Sintering of magnesium. Advanced Engineering Materials. 2010;**12**:829-836. DOI: 10.1002/adem.201000038

[59] Atrens A, Johnston S, Shi Z, Dargusch MS. Viewpoint-Understanding Mg corrosion in the body for biodegradable medical implants. Scripta Mater. 2018;**154**:92-100. DOI: 10.1016/j.scriptamat.2018.05.021

[60] Vaira Vignesh R, Padmanaban R, Govindaraju M. Investigations on the surface topography, corrosion behavior, and biocompatibility of friction stir processed magnesium alloy AZ91D. Surface Topography: Metrology and Properties. 2019;**7**:025020. DOI: 10.1088/2051-672X/ab269c

[61] Allavikutty R, Gupta P, Subhra Santra T, Rengaswamy J. Additive manufacturing of Mg alloys for biomedical applications: Current status and challenges. Current Opinion in Biomedical Engineering. 2021;**18**:100276. DOI: 10.1021/acsami.0c01497

[62] Gieseke M, Noelke C, Kaierle S, Wesling V, Haferkamp H. Selective laser melting of magnesium and magnesium alloys. In: Magnesium Technology. Vol. 2013. Hoboken, NJ, USA: John Wiley & Sons, Inc; 2013. pp. 65-68

[63] Niu X, Shen H, Fu J. Microstructure and mechanical properties of selective laser melted Mg-9 wt%Al powder mixture. Materials Letters. 2018;**221**:4-7. DOI: 10.1016/j.matlet.2018.03.068

[64] ASTM International, Standard Terminology for Additive Manufacturing. West Conshohocken, PA: ASTM International; 2015

[65] Karunakaran R, Ortgies S, Tamayol A, Bobaru F, Sealy MP. Additive manufacturing of magnesium alloys. Bioact Mater. 2020;**5**:44-54. DOI: 10.1016/j.bioactmat.2019.12.004

[66] Balog M, Snajdar M, Krizik P, Schauperl Z, Stanec Z, Catic A. Titanium-Magnesium Composite for Dental Implants (BIACOM). Pittsburg: 146th Annual Meeting & Exhibition of The Minerals, Metals & Materials Society, Springer Nature; 2017. pp. 271-284. DOI: 10.1007/978-3-319-51493-2

[67] Perets T, Ghedalia-Peled NB, Vago R, Goldman J, Shirizly A, Aghion E. In vitro behavior of bioactive hybrid implant composed of additively manufactured titanium alloy lattice infiltrated with Mg-based alloy. Materials Science & Engineering C. 2021;**129**:112418. DOI: 10.1016/j.msec.2021.112418

[68] Jiang G, Wang C, Li Q, Dong J, He G. Porous titanium with entangled structure filled with biodegradable magnesium for potential biomedical applications. Materials Science and Engineering: C. 2015;**47**:142-149. DOI: 10.1016/j.msec.2014.11.014

[69] Mohammed MT, Khan ZA, Siddiquee AN. Beta Titanium Alloys: The Lowest Elastic Modulus for Biomedical Applications: A Review. International Journal of Materials and Metallurgical Engineering. 2014;**8**:822-827. DOI: 10.5281/zenodo.1094481

[70] Torres Y, Trueba P, Pavón J, Montealegre I, Rodríguez-Ortiz JA. Designing, processing and characterisation of titanium cylinders with graded porosity: An alternative to stress-shielding solutions. Materials and Design. 2014;**63**:316-324. DOI: 10.1016/j.matdes.2014.06.012

[71] Mullen L, Stamp RC, Brooks WK, Jones E, Sutcliffe CJ. Selective laser melting: a regular unit cell approach for the manufacture of porous, titanium, bone in-growth constructs, suitable for orthopedic applications. J. Biomed. Mater. Res. - Part B Appl. Biomater. 2009;**89**:325-334. DOI: 10.1002/jbm. b.31219

[72] Zhang XZ, Leary M, Tang HP, Song T, Qian M. Selective electron beam

manufactured Ti-6Al-4V lattice structures for orthopedic implant applications: current status and outstanding challenges. Current Opinion in Solid State & Materials Science. 2018;**22**:75-99. DOI: 10.1016/j.cossms.2018.05.002

[73] Sochacka P, Miklaszewski A, Jurczyk M. Development of β-type Ti-x at. % Mo alloys by mechanical alloying and powder metallurgy: Phase evolution and mechanical properties ($10 \leq x \leq 35$). Journal of Alloys and Compounds. 2019;**776**:370-378. DOI: 10.1016/j.jallcom.2018.10.217

[74] Almeida A, Gupta D, Loable C, et al. Laser-assisted synthesis of Ti-Mo alloys for biomedical applications. Materials Science and Engineering: C. 2012;**32**:1190-1195. DOI: 10.1016/j.msec.2012.03.007

[75] Matsumoto H, Watanabe S, Hanada S. Beta TiNbSn alloys with low Young's modulus and high strength. Materials Transactions. 2005;**46**:1070-1078. DOI: 10.2320/matertrans.46.1070

[76] Bram M, Ebel T, Wolff M, Barbosa APC, Tuncer N. Applications of powder metallurgy in biomaterials. Advances in Powder Metallurgy. 2013;**2013**:520-554. DOI: 10.1533/9780857098900.4.520

[77] Eisenbarth E, Velten D, Müller M, et al. Biocompatibility of β-stabilizing elements of titanium alloys. Biomaterials. 2004;**25**:5705-5713. DOI: 10.1016/j.biomaterials.2004.01.021

[78] Cremasco A, Osório WR, Freire CMA, et al. Electrochemical corrosion behavior of a Ti-35Nb alloy for medical prostheses. Electrochimica Acta. 2008;**53**:4867-4874. DOI: 10.1016/j.electacta.2008.02.011

[79] Sumitomo T, Cáceres CH, Veidt M. The elastic modulus of cast Mg-Al-Zn alloys. Journal of Light Metals. 2002;**2**:49-56. DOI: 10.1016/S1471-5317(02)00013-5

[80] Arifvianto B, Zhou J. Fabrication of metallic biomedical scaffolds with the space holder method: A review. Materials (Basel). 2014;**7**:3588-3622. DOI: 10.3390/ma7053588

[81] Lutterotti L, Matthies S, Wenk HR. MAUD: a friendly Java program for material analysis using diffraction. IUCr: Newsletter of the CPD. 1999;**21**:14-15

[82] Burke P, Kipouros GJ, Fancelli D, et al. Sintering fundamentals of magnesium powders. Canadian Metallurgical Quarterly. 2009;**48**:123-132. DOI: 10.1179/cmq.2009.48.2.123

[83] Dutta G, Bose D. Effect of Sintering Temperature on Density, Porosity and Hardness of a Powder Metallurgy Component. Int J Emerg Technol Adv Eng. 2012;**2**:121-123

[84] Smorygo O, Marukovich A, Mikutski V, et al. High-porosity titanium foams by powder coated space holder compaction method. Materials Letters. 2012;**83**:17-19. DOI: 10.1016/j.matlet.2012.05.082

[85] Xu W, Liu Z, Lu X, et al. Porous Ti-10Mo alloy fabricated by powder metallurgy for promoting bone regeneration. Science China Materials. 2019;**62**:1053-1064. DOI: 10.1007/s40843-018-9394-9

[86] Chen XB, Li YC, Hodgson PD, et al. The importance of particle size in porous titanium and nonporous counterparts for surface energy and its impact on apatite formation. Acta Biomaterialia. 2009;**5**:2290-2302. DOI: 10.1016/j.actbio.2009.02.027

[87] Li JP, Li SH, Van Blitterswijk CA, et al. A novel porous Ti6A14V: Characterization

and cell attachment. J Biomed Mater Res - Part A. 2005;**73**:223-233. DOI: 10.1002/jbm.a.30278

[88] Wataha JC. Alloys for prosthodontic restorations. The Journal of Prosthetic Dentistry. 2002;**87**:351-363. DOI: 10.1067/mpr.2002.123817

[89] Low IM, Duraman N, Mahmood U. Mapping the structure, composition and mechanical properties of human teeth. Materials Science and Engineering: C. 2008;**28**:243-247. DOI: 10.1016/j.msec.2006.12.013

[90] Zysset PK, Guo XE, Ho CE, et al. Elastic modulus and hardness of cortical and trabecular bone lamellae measured by nanoindentation in the human femur. Journal of Biomechanics. 1999;**32**:1005-1012. DOI: 10.1016/s0021-9290(99)00111-6

[91] Dall'Ara E, Öhman C, Baleani M, et al. The effect of tissue condition and applied load on Vickers hardness of human trabecular bone. Journal of Biomechanics. 2007;**40**:3267-3270. DOI: 10.1016/j.jbiomech.2007.04.007

Chapter 4

The Role of Silane Sol-Gel Coatings on the Corrosion Protection of Magnesium Alloys

Emilia Merino, Alicia Durán and Yolanda Castro

Abstract

Magnesium alloys, as the lightest structural metallic material with promising physical, mechanical, and biodegradable properties, have become very attractive for different technical applications, especially for industrial and biomedical fields. However, rapid corrosion is the most critical obstacle that limits its use to play a major role in large-scale applications. The simplest way to control the corrosion rate is to prevent a direct contact of the magnesium substrate with the environment by using surface modification technologies. Silica sol-gel coatings are considered a promising solution to enhance the corrosion resistance of magnesium alloys because sol-gel-based coating systems form very stable chemical bonds with the metallic surface. In this chapter, an insight about the advances in silica sol-gel coatings as an alternative method to control the corrosion of Mg and its alloys will be exposed. A wide overview of the most relevant aspects and their current applications, specifically for aerospace, automobile, and biomedical applications will be described. The modification of silica sol-gel matrix by the incorporation of different types of inhibitors to achieve an active barrier property on Mg alloys has been also considered. Finally, the future perspective based on the development of new silica sol-gel coatings on Mg alloy will be presented.

Keywords: Mg alloys, sol-gel, protective coating, corrosion, synthesis sol-gel, inhibitor

1. Introduction

The use of magnesium alloys in different industrial fields has increased mainly due to its very high strength-to-weight ratio in comparison to other structural alloys [1]. However, an important limiting factor is their high reactivity and thus, their susceptibility to corrosion [2]. The main objective of the research area has always been to increase the corrosion resistance of metallic substrates [3]. One method to reduce the effect of corrosion is to deposit a protective coating on a metallic substrate. Among the coating techniques, sol-gel process is considered a very efficient and economically viable solution for developing anticorrosion coatings on magnesium alloys. According to Segal [4], the sol-gel process can be defined as the production of inorganic oxides in the form of colloidal dispersion or metal alkoxides.

The sol-gel process was initially developed for producing pure inorganic materials, ceramic, and glass materials. However, pure inorganic sol-gel coatings do not provide enough corrosion protection due to the presence of micro-cracks or defects [5]. The ability to process organic-inorganic hybrid composites at low temperature opened new opportunities in the design of free-crack sol-gel coatings that enhances the corrosion resistance of metals [6]. The research focused on the polymerization of organic-inorganic hybrid materials by sol-gel process increased significantly near the end of the twentieth century [7]. Thus, sol-gel process has got a strong technological impact on research related to protective and functional coatings because this method allows the surface modification of different materials without changing the substrate properties.

The citation report of the "Web of Science Core Collection" database reveals that the amount of literature containing "sol gel" and "Mg alloys" as keywords was 421 between 2000 and 2021; the research in this field is annually growing because of the new alkoxysilane precursors and functional species now available to obtain silica coatings with novel physicochemical properties.

1.1 A brief description of sol-gel synthesis process

The sol-gel technology is a wet-chemical process where the principal chemical aspect is the transformation of compounds, known as precursors, that contain Si-OR and Si-OH to form stable colloidal particle suspensions known as sol [8]. The sol can be applied on the substrates by different deposition techniques and then sintered to obtain a coating. During the aging step, a chemical transformation of the sol occurs leading to a rigid network, resulting in a gel [9]. Generally, inorganic or organic-inorganic sols are obtained via hydrolysis and polycondensation reactions between silicon alkoxides ($Si(OR)_4$) such as: tetramethoxysilane (TMOS) or tetraethoxysilane (TEOS) and organoalkoxysilanes R'-Si $(OR)_{n-1}$; where R' is the organic functional group linked to Si through a nonhydrolyzable covalent bond. During the hydrolysis stage, alkoxide groups are replaced with hydroxyl groups. Once the hydrolysis reaction has initiated, the condensation reaction occurs simultaneously. In this stage, the hydroxyl group and residual alkoxyl group react to form a three-dimensional Si−O−Si network [10]. From the reaction pathway point of view, two different Si−O−Si formation mechanisms can take place regarding if the reaction is performed under acidic or basic conditions. Therefore, the morphology and the structure of the resulting network strongly depend on the pH of the reaction.

Under acidic conditions [11], the oxygen atom of Si−O−R group is protonated in the first step to form a good leaving group. The central silicon atom turns to be more electrophilic and thus more susceptible to react by water to form Si−OH group (**Figure 1(1)**). An equilibrium condition is established between silanol groups and H^+ ions, resulting in positively charged species $Si{-}OH_2^+$ that interact with a silanol group to form Si−O−Si bonds (**Figure 1(2)**). In this case, the polymerization rate is directly proportional to the H^+ concentration. Therefore, a large number of monomers or small oligomers with reactive Si−OH groups are simultaneously obtained. The hydrolysis step reaction is favored, and the condensation step reaction is the rate-determining step. It was reported that the positively charged species, $Si{-}OH_2^+$, react preferentially with the less acidic silanols (silanols attached to the least condensed (Si−O−Si) end groups), giving to chain-like networks [13].

Under basic conditions [14], the hydrolysis reaction occurs directly by nucleophilic attack of OH^- to the silicon atom to form Si−OH bonds. In this case, deprotonated silanol ($Si{-}O^-$) anion is formed and then it gets condensed with a silanol group. The condensation reaction is favored, and the hydrolysis reaction is the rate-determining

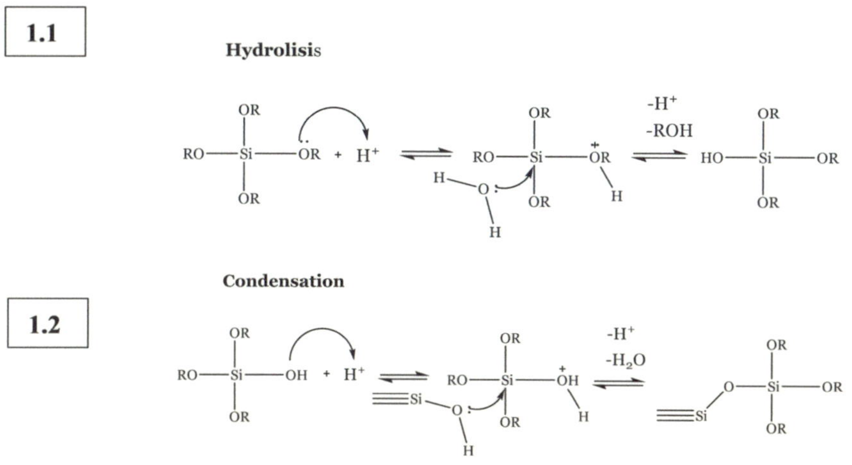

Figure 1.
Mechanism of acid-catalyzed sol-gel process. (1) Hydrolysis mechanism and (2) Condensation mechanism reactions. Image adapted from reference [12].

step. The hydrolyzed species are immediately consumed because of the fast condensation. Due to the nucleophilic nature of the deprotonated silanol, the $Si-O^-$ preferentially attacks the more acidic silanol (silanols attached to the highest condensed (Si−O−Si) end groups), leading to the formation of branched and highly condensed clusters.

1.2 Deposition methods and curing process of silane sol-gel coatings

A typical route of formation of silane sol-gel coating is described as the following process: synthesis of sol-gel > deposition > heat treatment. A sol-gel coating can be applied to Mg metallic substrates through various techniques, such as dip-coating, spin-coating, spraying, and electrodeposition, among others. However, dipping and spinning techniques are the two most used ones, especially for flat surfaces [10]. In the case of complex shapes, uniform coating can be obtained by electro-phoretic deposition method (EPD) [15].

By dip-coating, the surface treatment is attained by immersing the substrate into the sol-gel solution. The silanol groups Si−OH interact spontaneously with the Mg−OH groups that existed on the alloy surface via Van der Waals interactions. Upon the heat treatment, the Si−OH and Mg−OH bonds are attached firmly via a condensation reaction producing metallo-siloxane (Mg−O−Si) covalent bonds (**Figure 2**), and the remaining Si−OH groups of the deposited sol condense and form Si−O−Si bonds [16].

By controlling the curing temperature, the control of pore volume and size and mechanical strength can be achieved. High temperatures (more than 200°C) are normally used to cure inorganic sol-gel coatings and lower temperatures (less than 200°C) for drying/curing organic-inorganic sol-gel coatings [10]. Depending on

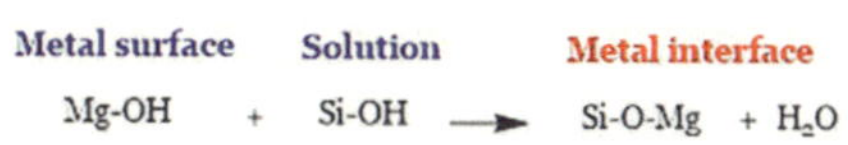

Figure 2.
Schematic representation of metallo-siloxane covalent bond formation.

the sol-gel precursors used, an optimal curing temperature should be defined since an inaccurate temperature could result in a decrement of the corrosion resistance properties of the coating and/or on the mechanical properties of the substrates. For instance, room temperature cured sol-gel coatings exhibit crack-free morphology, but a higher water sensitivity compared to coatings cured at a higher temperature. On the other hand, an increment of the curing temperature can lead to cracked coatings due to the stresses that appear during the sintering process [12]. A relatively new approach to densify sol-gel coatings is to use UV radiation [17]. Sol-gel films treated by UV radiation at room temperature can form denser sol-gel coatings able to improve the corrosion resistance of alloys.

1.3 Corrosion behavior of Mg alloys in aqueous environment

The poor corrosion resistance of Mg alloys can be mainly attributed to its high electronegative potential and the poorly protective properties of the quasi-passive oxide/hydroxide layer formed upon Mg. Generally, when the Mg alloys corrode in aqueous electrolyte, the metal changes its oxidation state, forms ionic species, and releases electrons. To maintain electroneutrality, the generated electrons must be consumed by other species. Therefore, the anodic reaction must be accompanied by a reduction reaction, where a molecule, ion, or atom gains electrons. In aqueous solution, water reduction is the dominant cathodic reaction. **Figure 3(1)** illustrates the anodic and cathodic reactions and the overall reaction that takes place during the corrosion of Mg in aqueous environment. The presence of chloride ions in the aqueous solution typically leads to accelerated corrosion processes (**Figure 3(1)**). $Mg(OH)_2$ can convert to $MgCl_2$, with higher solubility, promoting the dissolution of the Mg alloy [18].

On the other hand, the corrosive environment in the human body has a solution consisting of 0.14 M NaCl and other inorganic species, such as Ca^{2+}, PO_4^{3-}, and HCO_3^-. In this case, the presence of phosphates and carbonates promotes the formation of partially protective corrosion product layers [19]. It is clear that corrosion products depend on the type of the electrolyte. These corrosion products not only affect the corrosion rate but can also provide different protection properties to the substrate. **Figure 3(2)** shows a schematic representation of possible interactions between corrosion products of Mg alloy surface on a biological environment.

The deposition of a silane coating could control the corrosion of Mg alloys, although it could dissolve in contact with water due to the hydrolysis of the polysiloxane (Si−O−Si) network [20] that results in the release of silicic acid ($Si(OH)_4$), which can be expressed as follows [21]:

$$SiO_2(s) + 2H_2O \longleftrightarrow Si(OH)_4 \quad (1)$$

It is important to determine the corrosion rate to explain the corrosion behavior and provide models that predict the kinetics of the corrosion in an engineering context. The most widely used technique for exploring the corrosion behavior of a coated Mg alloy involves immersing the samples in a corrosive solution, since the corrosion performance is faster than atmospheric corrosion tests [19]. To study the corrosion performance of coated Mg alloys in aqueous solution, a wide range of tests are used [2]. These tests are divided into two large groups: electrochemical and nonelectrochemical

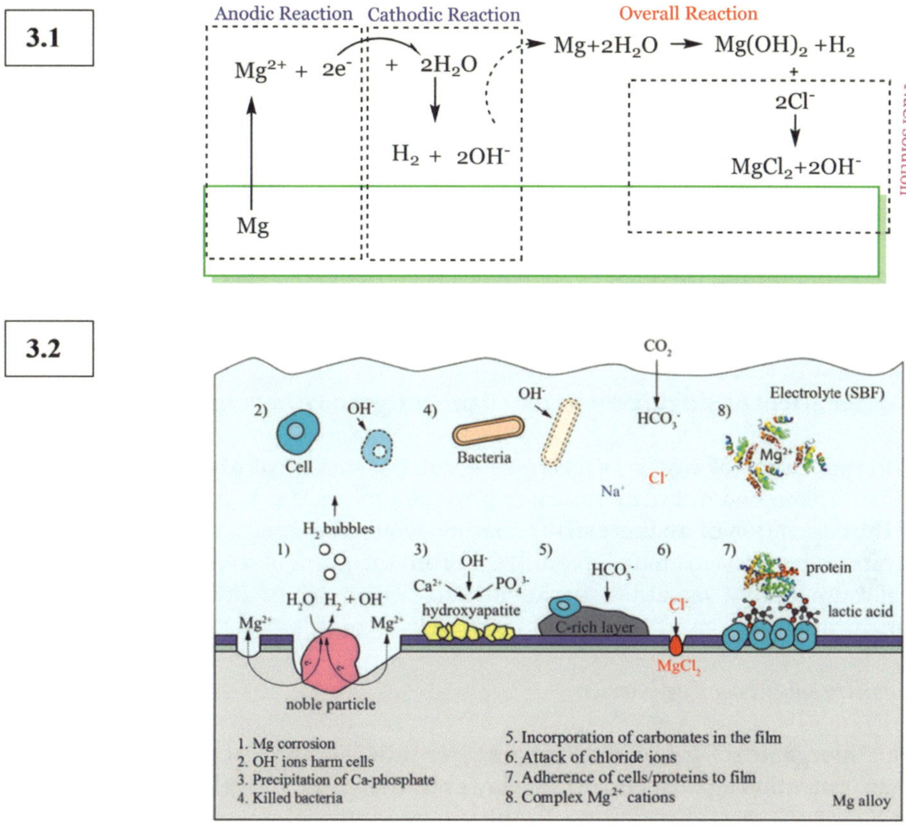

Figure 3.
Schematic representation of reactions that take place between Mg alloy surface and (1) NaCl aqueous solution and (2) biological environment (reprinted from Ref. [19], Elsevier).

tests [22]. The most common electrochemical and nonelectrochemical methods used are mentioned below.

1.3.1 Electrochemical techniques

These methods are important and rapid tools for assessing the corrosion of coated Mg alloys. Between the electrochemical techniques, the most used are Potentiodynamic polarization (PDP) and electrochemical impedance spectroscopy (EIS).

1.3.2 Nonelectrochemical techniques

The most common nonelectrochemical methods used for Mg corrosion research are weight loss measurements, hydrogen collection, and pH measurements.

The electrochemical techniques for exploring the corrosion behavior of coated Mg alloys can be used independently or simultaneously with the nonelectrochemical techniques. The principles of each technique and an overview of the main advantages and limitations of different techniques were provided by Durán et al. [5]. A recent review by Kirkland et al. [23] considers the methodologies used to study the corrosion of biodegradable Mg implant materials.

2. Potential use of the Mg alloys in industrial applications

One of the key techniques to reduce fuel consumption and subsequently greenhouse gas emission is to shift to lightweight vehicles. Magnesium alloys with their low density, easy recyclability, and high strength-to-weight ratio are exceptional candidates in the automotive and aerospace sectors [24]. However, the high chemical reactivity and the low standard corrosion potential (~−1.65 V.SCE), which is less electropositive than aluminum alloys (~ (−0.73 V.SCE)), make them highly susceptible to corrosion, limiting their use in such areas [25]. In order to prevent corrosion of Mg alloys, surface modification techniques such as sol-gel technology have attracted increasing interest for many researchers.

2.1 Barrier effect of single protective silane sol-gel coatings against corrosion

The application of single sol-gel coatings on the surface of Mg alloys has been considered as a good initial approach to provide a protective behavior to the alloy, since the deposition of an inert silane coating avoids the direct contact between substrate and corrosive environment [26]. From the point of view of synthesis, the sol-gel route offers a versatile way to synthesize effective and denser coatings with specific properties.

2.1.1 Passive sol-gel coating barrier

Pure inorganic sol-gel coatings have been studied as an inert physical barrier to provide protection against corrosion. However, inorganic sol-gel coatings have some limitations such as: (i) brittleness, shrinkage, and internal stress after heat treatment process, and (ii) the high temperature required to sinter the coating that mismatches with the thermal expansion coefficient of Mg substrate [27]. A great effort has been made to incorporate organo-alkoxysilanes into the sol-gel synthesis to obtain crack-free hybrid coatings able to be sintered at lower temperatures (below 200°C, depending on Mg alloys) close to the thermal expansion coefficient of Mg substrate [28]. Due to the wide variety of organic-inorganic precursors, there is growing attention on producing hybrid-inorganic sol-gel coatings with different cross-linked structures and compositions. The final hybrid sol-gel coatings could reach fascinating mechanical and physical properties such as flexibility, hydrophobicity, exceptional dielectric properties, strength, ductility, hardness, and good thermal stability. Zucchi et al. [29] studied the protective performance of coatings obtained using organo-silanes with a long alkyl chain (octadecyl-trimethoxysilane) on AZ31 magnesium alloy. An improvement of the corrosion resistance properties of Mg alloy was observed, confirming that the modification of the siloxane network using a long aliphatic chain provides a positive effect regarding corrosion performance.

The corrosion protection of organic-inorganic hybrid thin films prepared with other organoalkoxysilane precursors such as methacryloxypropyltrimethoxysilane (MAPTMS) and tetramethoxysilane (TMOS) on AZ31 and AZ61 Mg alloys has been also studied by El-Hadaba [30]. The results showed an enhancement of the corrosion protection properties at the initial immersion time, but a quick degradation of the coated AZ31 Mg alloy after 1 day of immersion in 0.6 M NaCl aqueous solution. This behavior was attributed to microscopic pores defects in the sol-gel layer. The low pH of the sol-gel solution promoted the Mg dissolution during the deposition process together with the hydrogen evolution during the curing sol-gel coating.

As observed, although sol-gel technology allows the preparation of different hybrid organic-inorganic sol-gel coating compositions, the obtention of effective coatings for Mg-based alloys still is a huge challenge. One of the main aspects, related to the synthesis of the sol-gel coatings on Mg alloys, is the pH of the hydrolyzed sol-gel solution. Indeed, magnesium is not stable and spontaneously degrades during sol-gel deposition step when using acidic conditions. Hernández-Barrios et al. [31] evaluated the corrosion behavior of AZ31 Mg alloy pretreated with a hybrid silica sol-gel coating prepared using acetic acid as acid-catalyst, and 3-glycidyloxypropyl-trimethoxysilane (GPTMS) and TEOS as silica precursors. The results revealed that the sol synthesized with the highest acid concentration reached more stable gelation kinetics, but with the worst corrosion resistance performance (i_{corr}: 1.3 × 10^{-6} A/cm^2) compared with the sol synthesized with the slower acid concentration (i_{corr}: 2.4 × 10^{-7} A/cm^2). The decay of the corrosion resistance of the sample coated with the more acidic sol is attributed to defects on the coating's morphology and to the corrosion process advancing in the substrate. Indeed, during the sol-gel deposition, corrosion products are generated together with hydrogen evolution.

In this sense, pH of the sol is a critical parameter and should be considered to get a nondefective SiO_2 coating not affecting the metallic substrate, and thus to provide a suitable corrosion resistance performance.

Another aspect to consider is related to the promotion of insulating coatings with high-density structures for blocking the penetration of electrolytes. In this case, complexing agents are added during the synthesis of the sol to react with the organic group of some organo-alkoxysilanes and therefore stimulate the organic polymerization. For instance, Qian et al. [32] prepared a hybrid sol-gel through hydrolysis and condensation reactions of TEOS and GPTMS. The opening of the epoxy group of GPTMS results in coatings with novel physical and chemical properties. The authors further incorporated triethylenetetramine (TETA) as an organic crosslinking agent to bond with the open epoxy groups. The corrosion behaviors of the coatings deposited on AZ31B magnesium alloy were evaluated by polarization curves measurements in the 3.5% NaCl solution. The results revealed that a compact and smooth silane film was formed on the substrate's surface, which provided good barrier protection, improving the corrosion resistance ability (i_{corr}: 3.7 × 10^{-9} A/cm^2) in comparison to untreated magnesium alloy substrate (i_{corr}: 4.1 × 10^{-6} A/cm^2) (**Figure 4**).

Furthermore, the corrosion resistance properties of silane films can be significantly improved by the incorporation of some nanoparticles into the sol-gel film. The beneficial effects of the addition of different nanoparticles on the corrosion resistance for Mg alloys have been reported by different researchers. For instance, the effect of incorporating SiO_2 nanoparticles [33], graphene oxide [34, 35], carbon nanotubes [36], alumina, titania, zirconia [37], and Montmorillonite (MMT) [38] on the sol-gel synthesis has been evaluated.

For example, the addition of a colloidal silica nanoparticles suspension into the sol-gel coating is considered a good approach to increase the hardness, density, and wear resistance, and thus the corrosion resistance properties of hybrid silane coatings. Peres et al. [33] investigated the effect of adding different amounts of SiO_2 nanoparticles into a hybrid silica sol based on TEOS and GPTMS on the corrosion resistance of AZ31 magnesium alloy. The results showed that the incorporation of nanoparticles improved the corrosion resistance of Mg alloy. However, the maximum amount of SiO_2 recommended to obtain a coating with the best anticorrosive performance was between 100 and 300 mg l^{-1}; coatings doped with a higher amount of SiO_2 showed nanoparticles agglomeration and consequent defects and cracks. Thus, two critical issues should be

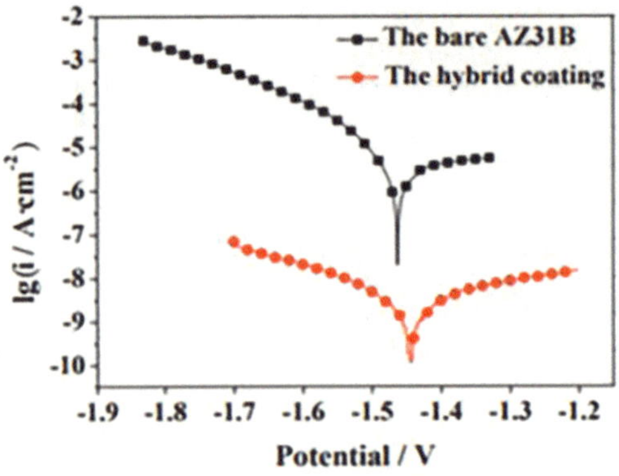

Figure 4.
Potentiodynamic polarization curves of bare alloy and the hybrid coating in 3.5 wt.% NaCl (reprinted from Ref. [32], ESG).

considered to avoid a detrimental effect on the anticorrosion behavior of the film: (i) the dispersion of nanoparticles into the film, and (ii) the amount of loaded nanoparticles.

On the other hand, graphene oxide (GO), which is a two-dimensional sp^2 carbon material, with many inherent characteristics such as good mechanical strength, chemical inertness, and good thermal stability, has also been considered to reinforce organofunctional silane coatings for corrosion protection of Mg alloys [34]. However, the high specific surface area of graphene and the strong Van der Waals force (−stacking) between graphene layers made it to agglomerate easily, resulting in hybrid coatings with a decrease in corrosion performance and microhardness properties. As graphene-based compound [39], oxidized fullerene [40], and carbon nanotube [36] have also been considered as a novel promising reinforcement for hybrid composite silane coatings for Mg alloys due to their properties including high strength, lightweight, thermal and mechanical stability, hydrophobicity, corrosion resistance, and high specific surface area. The anticorrosion and protective action of a zeolite-filled silane sol-gel coating on AZ31 magnesium substrate was studied by Calabrese et al. [41]. The zeolite composite coating evidenced very high hydrophobicity behavior (contact angle up to 140° showed good adhesion and good barrier properties during immersion in 3.5 wt.% NaCl solution.

Although significant advances have been made regarding modified sol-gel coatings composition, there is still a large gap in this research since most of the studies only provide information about the instantaneous corrosion rate, but not about the kinetic of the sol-gel film degradation in standard aqueous solution of 3.5 wt.% NaCl, which is helpful for a comprehensive choice of anticorrosion strategies and a systematic control of the degradation of sol-gel films.

2.1.2 Active sol-gel coatings barrier

Up to now, conventional physical barrier coatings with suitable composition designs have been considered to improve the corrosion resistance of Mg alloys; however, in very harsh environments when the aggressive agent and water reach the metal surface, the silane coatings are not capable to stop the corrosion process, reducing the lifetime of the coating protection. For this reason, smart self-healing protective coatings should be considered to provide long-term protection to the material. The smart self-healing effectiveness relies on a dissolving-reprecipitation interaction in the local defect, able to repair the defects entirely or partially, restoring the functionality of the coatings. The incorporation of corrosion inhibitors into the silane sol-gel coatings is the most studied strategy to obtain a self-healing ability of the silane coating thus enhancing the corrosion resistance of the metal.

Inorganic corrosion inhibitors such as rare earth inhibitors (cerium and lanthanum) have been demonstrated to be effective in the protection of magnesium alloy. For example, the rapid formation of oxygen vacancies in ceria lattice plays a crucial role in self-healing coating formulation since cerium cations interact with the OH^- ion released during the corrosion process, forming stable and insoluble cerium oxide/hydroxide species that precipitate in the surface and prevent further corrosion process. Prolonging the corrosion time, the deposited film gradually grows reducing the oxygen and electron transfer [42].

The effect of adding Ce and La salts as inhibitors as well as nanoparticles in silane solution has been explored as an opportunity and a challenge for researchers. Zanotto et al. [43] studied the corrosion resistance efficiency of 3-mercapto-propyl-trimethoxysilane (PropS-SH) coatings modified with cerium nitrate ($Ce(NO_3)_3 6H_2O$) deposited on AZ31 magnesium alloy. Moreover, Qiao et al. [44] studied the corrosion resistance behavior of 3-methacryloxypropyltrimethoxysilane coatings modified by lanthanum nitrate ($La(NO_3)_3{\cdot}6H_2O$) deposited on AZ31 Mg alloy. Both studies demonstrated that either cerium or lanthanum ions can be added as inhibitors to the silane solutions to enhance the corrosion of the pretreatments for magnesium alloy. However, they reported that silane coatings doped with cerium nitrate salt showed poorer corrosion behavior than those doped with cerium nanoparticles, CeO_2NPs. Coatings doped with CeO_2NPs were more "compact," avoiding the electrolyte penetration, and therefore providing improved corrosion protection [45]. Under this perspective, Calado et al. [46] modified a hybrid epoxy-silane coating with ceria nanoparticles to improve the barrier protection of AZ31 Mg alloys. EIS results showed an improvement in corrosion resistance because the modified ceria-coating was capable to provide active corrosion protection. The ceria nanoparticles react with water and/or hydroxyl ions, producing a cerium (IV) oxide or hydroxide layer onto the AZ31 surface. Electrolyte diffusion pathways are blocked; thus, the localized corrosion activity is reduced.

The use of organic compounds with heteroatoms such as N, S, and O can provide inhibitory effects to silane coatings. The major role of heteroatoms in corrosion protection is the formation of a complex chelate with Mg^{2+} ions which create insoluble deposits on the metallic surface, blocking the active sites and preventing the local pH increases, which is responsible for the intensification of intermetallic dealloying [47]. Toorani et al. [48] proposed a silane coating with active corrosion properties using γ-amino propyltriethoxysilane (APS) and TEOS as silica precursors, and adding different organic inhibitors: 8-hydroxyquinoline (8-HQ), indole-3-carbadehyde (I3C), 2-mercaptobenzoxazole (MBO), and sodium diethyldithiocarbamate (DDTC) to silane precursors. The results showed that organic inhibitors provide better active corrosion protection properties to the silane coating compared to the bare AZ91D magnesium alloy, especially when the 8-HQ inhibitor was added. In search of new organic inhibitors for corrosion protection of Mg alloys, Ashassi-Sorkhabi et al. [49] reported the effect of adding amino acids (L-alanine, L-glutamine, L-methionine, and L-aspartic amino acids) as eco-friendly inhibitors into sol-gel coating matrix. The corrosion ability of amino acids was associated with their tendency to form hydrogen bonds with the oxide or hydroxide groups on the metal surface and to the lone pair electrons present in their heteroatoms that can complex Mg cation. The paper described that all amino acids improved the anticorrosion performance of the silane coating, but L-aspartic exhibited the best enhancement effect.

2.2 Barrier effect of multilayer protective coatings

Even though silica sol-gel coatings have shown to be successful as a physical and active barrier, it is sometimes not enough for a long-term protection system in harsh environment. Some micro-defects or micro-cracks appear, allowing the penetration of corrosive agents and producing oxide-hydroxide-carbonate deposits beneath the coating, causing its rapid delamination. Thereby, the application of single-layer coating does not provide a full protection of Mg alloys. For this reason, great effort is underway to identify efficient alternative systems with desirable surface properties. In this context, the combination of different systems has been suggested, based on the deposition of a first oxide layer using conventional anodization or plasma electrolyte oxidation (PEO) processes followed by the deposition of silica sol-gel coating seems to be a good alternative.

PEO is an electrochemical process that has increasingly been employed to improve the surface properties of Mg alloys. This process produces an adhesive micro-porous oxide layer on the surface that provides a moderate protection on the metal and alloys. The ceramic-like film can be sealed with a silane coating to reduce the infiltration of the aggressive medium through the micro-pores, providing a long-term corrosion protection. Tan et al. [50] reported the preparation of a multilayer system obtained by anodizing the AZ91D Mg alloy and post-sol-gel treatment using MEMO (3-methacryloxypropyl trimethoxysilane), TPTMS (3-mercaptopropyl trimethoxysilane), and silica nanoparticles as reinforcement. The preliminary results showed that after the deposition of various silane layers by spray method, the silane coatings seal the pores of the anodized coating providing a physical corrosion protection in 3.0 wt.% NaCl.

Recently, Merino et al. [51] studied the corrosion resistant of an integrated system for AZ31B Mg alloys combining PEO and sol-gel process. In this case, the sol was prepared by using TEOS, GPTMS, colloidal SiO_2 nanoparticles, and 1-methylimidazole (MI), and then deposited onto optimized oxide coating. The results revealed that the multilayer system exhibits a good corrosion performance in 3.5 wt.% NaCl, since the polarization resistance (Rp) for the integrated system samples showed a quite high value (31546.8 Ω cm^2) compared to Mg alloy (207.3 Ω cm^2) (**Figure 5**).

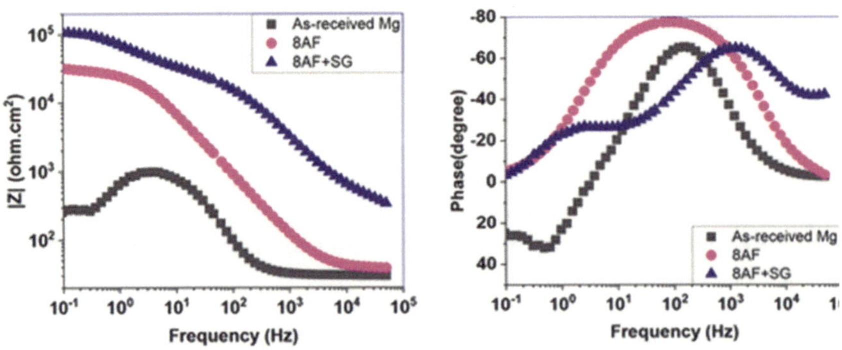

Figure 5.
Bode plot and phase angle plot for bare AZ31B Mg alloy, anodized sample and multilayer system tested in 3.5 wt.% NaCl (reprinted with permission from Ref. [51], Wiley Online Library).

This is an interesting alternative to significantly improve the corrosion resistance of Mg alloys. However, only a few papers present complete and decisive results. Additionally, different factors need to be considered to reach a good compromise between stacking and anticorrosion properties, such as sealing pore effectiveness and micro-cracks formation during the deposition of multiple silane layers [52].

3. Bio-applications

Mg alloys are considered as suitable candidate materials for biomedical applications due to their mechanical properties and their confirmed biocompatibility. In biological environments, magnesium alloys biodegrade with kinetics that depend on the surrounding tissue, eliminating secondary surgical procedure of implant removal [53]. The desirable Young's modulus of Mg alloys (41–45 GPa), which is close to the cortical bone (3–20 GPa), and the excellent ability of the Mg ions to promote bone regeneration make them attractive as orthopedic implants [54]. Currently, researchers are underway to improve bioresorbable cardiovascular stents based on Mg alloys, which are designed to provide short-term supporting structures and to combat coronary heart and peripheral artery diseases [55].

Soluble magnesium ions (Mg^{2+}), hydroxide ions (OH^-), and hydrogen gas (H_2) are well known for being the primary magnesium corrosion products. Many studies have confirmed that Mg^{2+} ions are essential for living cells and the excess can be excreted in the urine without causing damage to excretory organs such as the liver or the kidney. However, the rapid corrosion rate of Mg-based alloys in physiological conditions promotes an intense hydrogen evolution [56]. Hydrogen gas is nontoxic and is easily diffusible, but excessive corrosion leads to the formation of undesirable gas bubbles (emphysema) in surrounding soft tissue. The rapid evolution of H_2 bubbles can get accumulated and form gas pockets, leading to intensifying necrosis and inflammation within the living tissues [57]. On the other hand, depending on the type of the implant, the excessive corrosion leads to secondary problems. In the case of orthopedic implants, an excessive corrosion can produce early losing mechanical strength properties avoiding the implant assist the fracture of the bone firmly at least in the early healing stages (typically 12 weeks) [58]. Moreover, the uncontrollable and uneven degradation behavior for a vascular implant will produce huge amounts of hydrogen within a short time disfavoring the healing of neovascularization tissues which easily result in restenosis. Studies have shown that the critical period of vascular healing normally ends 3 months after implantation [58].

Both orthopedic and vascular magnesium implants look promising, but these drawbacks limit their applications. Thus, the use of Mg alloys as biodegradable implants is still in its infancy due to its high susceptibility to corrosion.

3.1 The effect of a silane coatings to control Mg alloy degradation

Since, silane coatings have demonstrated excellent biocompatibility, favorable cellular adhesion, and proper protein absorption, they have been employed as biofunctional coatings to control the high in-vivo degradation rate of Mg and its alloys. It has also been reported that organofunctional silane coatings do not cause adverse tissue reactions, and the degradation product ($Si(OH)_4$) produced into the body can be easily eliminated through the renal system. For this reason, some researchers have developed different compositions of organo-inorganic silane coatings for this application. For instance, Gaur et al. [59] studied the effect of a phosphonatosilane coating, trying to improve the corrosion resistance of Mg-6Zn-Ca magnesium alloy in a physiological environment. In this study, the authors used a phosphonate (silane diethylphosphatoethyltriethoxysilane (DEPETES)) and bis sulfur silane (bis-[3-(triethoxysilyl) propyl] tetrasulfide (BTESPT)) precursors to synthetize the silane coating, considering that both precursors were found to be nontoxic. The in-vitro investigation showed that the silane coating provided significant and durable corrosion resistance. Moreover, the presence of hydrated magnesium phosphate was also identified after 216 h of immersion test in m-SBF; component reported to support osteoblast formation and tissue healing. Two years later, the same authors [60] reported the preparation of other silane coating composition obtained by using GPTMS and MTEOS to improve the in-vitro corrosion resistance and biocompatibility of the Mg6ZnCa alloy. The results demonstrated that the deposition of a silica coating obtained by combining both precursors slow down the dissolution of a biodegradable magnesium alloy in the early stages (280 h), enhancing cells growth on the coated specimen. Furthermore, the formation of magnesium/calcium phosphate on the surface of the Mg alloys after immersion time showed good bioactivity and osteo conductivity of the coating. The results suggested that the sol-gel coating developed for the Mg6ZnCa alloy is a promising solution for biomedical application such as bio-absorbable surgical skin staples (needs to be removed after 10–12 days of postsurgery), micro-clips (needs to degrade within 2 weeks), and pins used in fingers dislocation or fracture that are predicted to heal quickly.

To enhance the corrosion resistance of magnesium alloys, a modified epoxy-silane coating obtained by using GPTMS and diethylenetriamine (DETA) as organic cross-linker was also proposed by Zomorodian et al. [61]. Although hydrogen evolution, pH, and in-vitro cell culture tests were not carried out, the EIS data showed an improvement of the corrosion resistance properties in Hank's solution associated with a dense and homogenous coating deposited on the Mg alloy. On the other hand, Castro et al. [62] also investigated the corrosion degradation rate of Mg alloys (AZ31B and AZ91D) by the deposition of two different silica sols prepared with and without colloidal silica particles for biodegradable implant materials. The results showed that the corrosion resistance behavior of Mg alloys, characterized in SBF using three different in-vitro tests: hydrogen evolution, pH variation, and potentiodynamic curves, enhanced after the deposition of the silane coating that contains nanoparticles as cross-linked network reinforcement (**Figure 6**).

Recent studies consider the development of double nano-composite coatings [63], based on the first deposition of $Mg(OH)_2$ or MgO enriched oxide layer and a subsequent deposition of a silane sol-gel coating, to achieve longer corrosion protection systems. Dou et al. [64] prepared double composite coatings using a conventional micro-arc oxidation process, and then the sol-gel technique. The in-vitro degradation performance of the composite coatings showed an improvement of the corrosion resistance properties by reducing the corrosion current density.

Different approaches have been considered to improve the biocorrosion resistance of the Mg alloys for cardiovascular stent application since it is a disease with high mortality and an increasing incidence [65]. For example, Liu et al. [66] reported the use of layer-by-layer self-assembly technique, based on the deposition of a first APTES-based silane coating followed by the deposition of a graphene oxide (GO) suspension. The results showed that the silane/GO composite coating improves the corrosion and wear resistance of Mg alloy, suggesting its use in biomedical fields as a vascular stent.

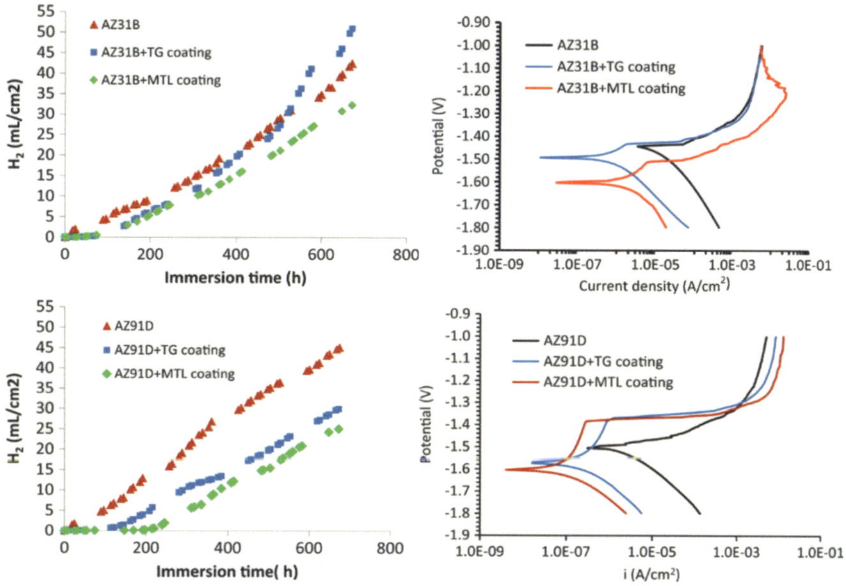

Figure 6.
Variation of hydrogen evolution as a function of immersion time in SBF solution for coated and uncoated AZ31B and AZ91D substrates; MTL coating corresponds to the silane coating that contains nanoparticles and TG to the silane coating without nanoparticles. (reprinted with permission from Ref. [62], Springer Nature).

3.2 Factors affecting the bio-functionality of silane coatings

3.2.1 Cyto-compatibility

The biocompatibility of Mg alloys is determined by the toxicity of the released corrosion products and the interaction effect between metal surface and living tissues. Not all the studies mentioned in the previous section included in-vitro cell viability tests as complementary information, necessary to determine the response of the silane coatings deposited on Mg alloys.

Although AZ91D showed a better corrosion resistance performance with respect to AZ31B Mg alloys, the AZ91D shows lower biocompatibility and bioactivity due to its higher Al content. To improve the cyto-compatibility of AZ91 Mg alloys, Witecka et al. [67] studied the effect of the deposition of different silane coatings on its surface; ethyltriethoxysilane (S_1), 3-aminopropyltriethoxysilane (S_2), 3-isocyanatopyl-triethoxysilane (S_3), phenyltriethoxysilane (S_4), and octadecyltriethoxysilane (S_5). S_1 was used to introduce a simple polysiloxane precursor to the substrate; S_2 and S_3 were selected to introduce positive and negative charges on the alloy surface, and finally, S_4 and S_5 were chosen to examine the π electrons and the long alkyl chain effect on the surface. Cell culture experiments showed that the cyto-compatibility was not affected by the surface modification. However, Silane S_1 was the only system able to improve cell growth during 7 days of incubation. Because cyto-compatibility is a basic and important parameter in the design of silane coating to bio application, other strategies have been considered to improve the bioactivity of Mg alloys, the deposition of sol-gel derived bioactive glasses coatings being one of them. These coatings based on pure silica, SiO_2-CaO-P_2O_5 or SiO_2-CaO, have been shown the largest level of bioactivity based on their reaction rate and bone binding ability. The bone-bonding ability occurs through the development of a biological apatite layer when the materials are exposed

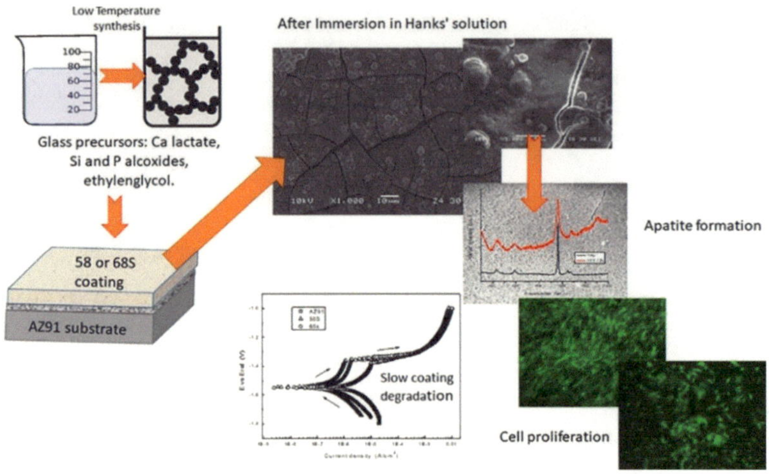

Figure 7.
Schematic representation of the deposition sol-gel glass-like bioactive sol for enhancing the implant performance of AZ91D magnesium alloy (reprinted with permission from [70]; Elsevier).

to body fluids or simulated body fluids. In-vivo studies have shown that biological species are incorporated into the silica-rich and apatite layers. Consequently, coatings react with the physiological fluid for obtaining an adequate interfacial bonding with bone by forming hydroxyapatite layer (HA). Their main applications are focused on bone repair and regeneration in the field of tissue engineering. Regarding the synthesis, some attempts have been made to obtain bio-glass silane coatings [68, 69]. Recently, Omar et al. [70] synthesized two compositions of bioactive silica-glasses, 58S and 68S, by using tetramethyl orthosilicate (TMOS), methyltriethoxysilane (MTES) and calcium L-lactate hydrate (**Figure 7**).

The lactate was used to avoid the use of calcium nitrate as a precursor due to the presence of nitrate residuals in the coatings is not beneficial to the body. The results showed that both coatings showed a quick apatite formation, good corrosion resistance properties, good cell adhesion, and proliferation, representing a promising coating system for degradable AZ91D implants.

Other strategies considered in the synthesis of silane-coatings to potentially improve the biocompatibility of Mg alloys consisted of the incorporation of hydroxyapatite nanoparticles. Nikbakht et al. [71] synthetized a modified silane coating with hydroxyapatite nanoparticles to promote biocompatibility and bone healing through producing calcium phosphorus-rich corrosion products. The results showed that a correct amount of hydroxyapatite nanoparticles not only helped to optimize the barrier properties of the silane coating, but also improved cell growth, especially the MG-63 osteoblastic.

3.2.2 Protein absorption-platelet adhesion

The initial interaction of biomaterials with the biological environment is based on the absorption of protein on the surface and the interaction with ions and water molecules to form various reactive interfaces. Understanding protein adsorption mechanisms, kinetics, and thermodynamics are essential to improve the design of silane biocompatible coatings [72]. Appropriate protein adsorption on the modified Mg surface alloys is essential for application in bone tissue regeneration and the effective integration of Mg implants. In relation to biomedical implants, such as cardiovascular stents, protein adsorption kinetics play a crucial role in the platelet adhesion process. Implanted biomaterials in contact with blood in vivo need to retain a low degree of platelet adhesion to prevent thrombosis, implant failure, and other complications [73]. A rapid adsorption of proteins might cause a higher number of platelet adhesion on the surface, which can trigger thrombus formation by platelet activation and ultimately result in blood coagulation. Considering that the protein adsorption can be roughly controlled through roughness and wettability, Majumder et al. [74] proposed the deposition of a hydrophobic silane-PMMA coating to improve the corrosion resistance and the hemo-compatibility nature of AE42 Mg alloys for cardiovascular stent applications. The results showed that an improvement in hydrophobicity resulted in a significant reduction of protein adsorption and hemolysis ratio, making it a favorable candidate for biodegradable stent application. Surface modification by the addition of Heparin (anticoagulant reagent) has also been considered to increase the thromboresistance of biomedical implants. Liu et al. [73] developed a biofunctionalized anticorrosive coating on Mg AZ31 alloy containing heparin reagent. The modified silane coating system reduced platelet adhesion on the surface, thus increasing its interest in biodegradable implant applications as cardiovascular stents.

3.2.3 Drug release silane coatings

On the other hand, bacterial infections or inflammations are one of the reasons for biomedical implants failure. Bacteria can form recalcitrant biofilms on implant surfaces, resisting conventional antibiotic treatments. As a consequence, the entire implant must be removed to allow an efficacious antibiotic treatment. Thus, it is necessary to find an effective local drug-releasing coating to simultaneously provide high anticorrosion and antibacterial ability for Mg alloys. Sol-gel coatings have attracted great attention since they offer the possibility to introduce antibiotics in the coating, and also to control the mechanism and kinetics of the drug release. Under this context, Xue et al. [75] designed a composite coating on AZ31 Mg alloy by depositing a drug-loaded coating obtained by crosslinking ciprofloxacin (CIP) (antibacterial drug) and polymethyltrimethoxysilane (PMTMS) as precursors. Cyto-compatibility and antibacterial performance of the coating were probed using in-vitro cytotoxicity tests (MTT), live/dead cell staining, and plate counting method. The results showed that the coating displayed a controllable long-term drug release ability against *Staphylococcus aureus* and *Escherichia coli*, implying that this coating provides a new approach for the anti-infection Mg bone implants. Although this coating system showed a good antibacterial performance, the drug-releasing time of the coating was too long to be used in the human body, since the period in the wound healing cycle is normally 1–2 weeks.

This approach can be a promising alternative, but it is necessary to continue studying ways to shorten drug-release time by modifying the synthesis of the sol. The control of the sol-gel synthesis and processing parameters together with the selection of the precursors are key issues.

4. Conclusions and perspectives

This chapter summarizes the advances of the silica sol-gel coating as a surface modification technique to control the corrosion of Mg and its alloys. The most important advantages of the sol-gel technique are the opportunity to introduce a wide range of alkoxysilane precursors and organic molecules in the synthesis for obtaining hybrid organic-inorganic sol-gel coatings with desirable cross-linking structure and good protective corrosion behavior. The organically modified sol-gel coatings provide the possibility to obtain thick, crack-free coatings with good corrosion performance. The hybrid films can be reinforced by doping with nanoparticles to obtain denser coatings, and with inhibitors to obtain active barrier protection. Although relevant advances have been made in recent years, some aspects related to the sol-gel technique on Mg alloys should be considered before obtaining a successful industrial application, especially for aerospace, automobile, and biomedical applications.

Since the corrosion behavior of sol-gel coatings depends on the synthesis parameters, organic-inorganic precursors, and the mechanical and chemical features of the comprising organic and inorganic networks, a variety of sol-gel coating with different compositions and cross-linked structures have been developed. However, the different protective properties and the service life between those coating on Mg alloys are still not known clearly. Therefore, systematic and long-term comparisons need to be conducted in future research to better understand the corrosion mechanism, as well as the advantages and disadvantages of each coating. Furthermore, the kinetics of hydrolysis and condensation reactions, gelation kinetics, and curing process parameter should also be studied and considered to avoid cracks coating formation during the heat post-treatment. Sol-gel films treated by UV radiation at room temperature can form a denser sol-gel coating that can improve the corrosion resistance of alloys.

Although a sol-gel coating is a promising alternative, recent works show that the deposition of a single layer of sol-gel coating faces many difficulties and does not stop the corrosion of Mg alloys. On that basis, the combination of different deposition processes, such as anodization or PEO processes, and sol-gel technique could be more effective methods to mitigate the corrosion damage. However, different factors such as sealing pore effectiveness should be considered to reach a good compromise between stacking and anticorrosion properties. The preparation of efficient composite coatings for Mg-based alloys is still a huge challenge.

To achieve a practical application in the biomedical field, the design of sol-gel coatings should be more purposeful. For example, for bone implant applications, the hybrid coatings should be pro-osteogenesis and biocompatible. Moreover, the corrosion resistance studies of the silane coatings deposited on Mg alloys should be complemented with in-vitro cell viability tests to determine the bifunctionality response of the silane coatings.

However, many challenges need to be faced and solved, intelligent multilayer systems are promising alternatives to significantly increase the use of Mg alloys in many relevant applications, from corrosion protection to bioactive devices. Continues research is the best way to get them.

Acknowledgements

This chapter is a part of the dissemination activities of the project FunGlass, which has received funding from the European Union's Horizon 2020 research and innovation program under grant agreement number: 739566.

Author details

Emilia Merino, Alicia Durán and Yolanda Castro*
Institute of Ceramics and Glass (CSIC), Madrid, Spain

*Address all correspondence to: castro@icv.csic.es

References

[1] Yang Y, Xiong X, Chen J, Peng X, Chen D, Pan F. Research advances in magnesium and magnesium alloys worldwide in 2020. Journal of Magnesium and Alloys. 2021;**9**(3):705-747

[2] Atrens A, Shi Z, Mehreen SU, Johnston S, Song GL, Chen X, et al. Review of Mg alloy corrosion rates. Journal of Magnesium and Alloys. 2020;**8**(4):989-998

[3] Figueira RB, Silva CJR, Pereira EV. Organic–inorganic hybrid sol–gel coatings for metal corrosion protection: A review of recent progress. Journal of Coating Technology and Research. 2015;**12**(1):1-35

[4] Segal DL. Sol-gel processing: Routes to oxide ceramics using colloidal dispersions of hydrous oxides and alkoxide intermediates. Journal of Non-Crystalline Solids. 1984;**63**(1-2):183-191

[5] Durán A, Castro Y, Conde A, Damborenea JJ. Sol-gel protective coatings for metals. In: Klein L, Aparicio M, Jitianu A, editors. Handbook of Sol-Gel Science and Technology: Processing, Characterization and Applications. Cham: Springer; 2018. pp. 2369-2433

[6] Wen J, Wilkes GL. Organic/inorganic hybrid network materials by the sol-gel approach. Chemistry of Materials. 1996;**8**(8):1667-1681

[7] Li Q. Sol-gel coatings to improve the corrosion resistance of magnesium (Mg) alloys. In: Song G-L, editor. Handbook of Corrosion Prevention of Magnesium Alloys. Sawston: Elsevier, Woodhead Publishing; 2013. pp. 469-485

[8] Hench LL, West JK. The sol-gel process. Chemical Reviews. 1990;**90**(1):33-72

[9] Pierre AC. Introduction to Sol-Gel Processing. 2nd ed. Boston, MA: Springer; 2020

[10] Mitzi DB. Thin-film deposition of organic-inorganic hybrid materials. Chemistry of Materials. 2001;**13**(10):3283-3298

[11] Levy D, Zayat M. The Sol-gel Handbook: Synthesis, Characterization, and Application. New York: John Wiley & Sons; 2015

[12] Gonzalez E, Vejar N, Solis R, Muñoz L, Encinas MV, Paez M. Sol-gel films: Corrosion protection coating for aluminium alloy. In: Aguilar GV, editor. handbook of Sol-Gel Method – Design and Synthesis of New materials with Interesting Physical. London, UK: Intechopen; 2019

[13] Zerda TW, Artaki I, Jonas J. Study of polymerization processes in acid and base catalyzed silica sol-gels. Journal of Non-Crystalline Solids. 1986;**81**(3):365-379

[14] Issa AA, Luyt AS. Kinetics of alkoxysilanes and organoalkoxysilanes polymerization: A review. Polymers. 2019;**11**(3):537

[15] Saji VS. Electrophoretic (EPD) coatings for magnesium alloys. Journal of Industrial and Engineering Chemistry. 2021;**103**:358-372

[16] Wang D, Bierwagen GP. Sol-gel coatings on metals for corrosion protection. Progress in Organic Coating. 2009;**64**(4):327-338

[17] Liu F, Liu A, Tao W, Yang Y. Preparation of UV curable organic/ inorganic hybrid coatings—A

review. Progress in Organic Coating. 2020;**145**:105685

[18] Lei T, Ouyang C, Tang W, Li LF, Zhou LS. Preparation of MgO coatings on magnesium alloys for corrosion protection. Surface and Coatings Technology. 2010;**204**(23): 3798-3803

[19] Esmaily M, Svensson JE, Fajardo S, Birbilis N, Frankel GS, Virtanen S, et al. Fundamentals and advances in magnesium alloy corrosion. Progress in Materials Science. 2017;**89**:92-193

[20] Toirac B, Garcia-Casas A, Cifuentes SC, Aguilera-Correa JJ, Esteban J, Mediero A, et al. Electrochemical characterization of coatings for local prevention of Candida infections on titanium-based biomaterials. Progress in Organic Coating. 2020;**146**:105681

[21] Juan-Díaz MJ, Martínez-Ibáñez M, Hernández-Escolano M, Cabedo L, Izquierdo R, Suay J, et al. Study of the degradation of hybrid sol-gel coatings in aqueous medium. Progress in Organic Coating. 2014;**77**(11):1799-1806

[22] Atrens A, Song GL, Shi Z, Soltan A, Johnston S, Dargusch MS. Understanding the corrosion of mg and mg alloys. In: Reedijk J, Kakeya H, Lammertsma K, Marquardt R, et al., editors. Handbook of Reference Module in Chemistry, Molecular Sciences and Chemical Engineering. Amsterdam: Elsevier; 2017. pp. 1-20

[23] Kirkland NT, Birbilis N, Staiger MP. Assessing the corrosion of biodegradable magnesium implants: A critical review of current methodologies and their limitations. Acta Biomaterialia. 2012;**8**(3):925-936

[24] Abela S. Protective coatings for magnesium alloys. magnes alloy. In: Czerwinski F, editor. Handbook of Magnesium Alloys Corrosion and Surface Treatments. London, UK: Intechopen; 2011

[25] Baillio SS. Corrosion protection of aerospace grade magnesium alloy Elektron 43^{TM} for use in aircraft cabin interior [master's degree]. Ann Arbor: University of North Texas; 2013

[26] Zhang D, Peng F, Liu X. Protection of magnesium alloys: From physical barrier coating to smart self-healing coating. Journal of Alloys and Compounds. 2021;**853**:157010

[27] Talha M, Ma Y, Xu M, Wang Q, Lin Y, Kong X. Recent advancements in corrosion protection of magnesium alloys by silane-based sol-gel coatings. Industrial and Engineering Chemistry Research. 2020;**59**(45):19840-19857

[28] Prasad SVS, Prasad SB, Verma K, Mishra RK, Kumar V, Singh S. The role and significance of Magnesium in modern day research—A review. Journal of Magnesium and Alloys. 2021. DOI: 10.1016/j.jma.2021.05.012

[29] Zucchi F, Frignani A, Grassi V, Balbo A, Trabanelli G. Organo-silane coatings for AZ31 magnesium alloy corrosion protection. Materials Science. 2008;**110**(2-3):263-268

[30] El-Hadad AA, Barranco V, Samaniego A, Llorente I, García-Galván FR, et al. Influence of substrate composition on corrosion protection of sol-gel thin films on magnesium alloys in 0.6M NaCl aqueous solution. Progress in Organic Coating. 2014;**77**(11):1642-1652

[31] Hernández-Barrios CA, Cuao CA, Jaimes MA, Coy AE, Viejo F. Effect of the catalyst concentration, the immersion time and the aging time on the morphology, composition and

corrosion performance of TEOS-GPTMS sol-gel coatings deposited on the AZ31 magnesium alloy. Surface and Coating Technology. 2017;**325**:257-269

[32] Qian Z, Wang S, Wu Z. Corrosion behavior study of AZ31B magnesium alloy by sol-gel silica-based hybrid coating. International Journal of Electrochemical Science. 2017;**12**(9):8269-8279

[33] Peres RN, Cardoso ESF, Montemor MF, de Melo HG, Benedetti AV, Suegama PH. Influence of the addition of SiO_2 nanoparticles to a hybrid coating applied on an AZ31 alloy for early corrosion protection. Surface and Coatings Technology. 2016;**303**(Part B):372-384

[34] Li J, Cui J, Yang J, Ma Y, Qiu H, Yang J. Silanized graphene oxide reinforced organofunctional silane composite coatings for corrosion protection. Progress in Organic Coating. 2016;**99**:443-451

[35] Malik MU, Tabish M, Yasin G, Anjum MJ, Jameel S, Tang Y, et al. Electroless codeposition of GO incorporated silane nanocomposite coating onto AZ91 Mg alloy: Effect of GO content on its morphology, mechanical and corrosion protection properties. Journal of Alloys and Compounds. 2021;**883**:160790

[36] Nezamdoust S, Seifzadeh D, Rajabalizadeh Z. PTMS/OH-MWCNT sol-gel nanocomposite for corrosion protection of magnesium alloy. Surface and Coating Technology. 2018;**335**:228-240

[37] Zhu R, Zhang J, Chang C, Gao S, Ni N. Effect of silane and zirconia on the thermal property of cathodic electrophoretic coating on AZ31 magnesium alloy. Journal of Magnesium and Alloys. 2013;**1**(3):235-241

[38] Ashassi-Sorkhabi H, Moradi-Alavian S, Jafari R, Kazempour A, Asgharo E. Effect of amino acids and montmorillonite nanoparticles on improving the corrosion protection characteristics of hybrid sol-gel coating applied on AZ91 Mg alloy. Materials Chemistry and Physics. 2019;**225**:298-308

[39] Afsharimani N, Talimian A, Merino E, Durán A, Castro Y, Galusek D. Improving corrosion protection of Mg alloys (AZ31B) using graphene-based hybrid coatings. International Journal of Applied Glass Science. 2021;**13**(1):143-150

[40] Samadianfard R, Seifzadeh D, Habibi-Yangjeh A, Jafari-Tarzanagh Y. Oxidized fullerene/sol-gel nanocomposite for corrosion protection of AM60B magnesium alloy. Surface and Coatings Technology. 2020;**385**:125400

[41] Calabrese L, Caprì A, Proverbio E. Anti-corrosion performances of hybrid silane coatings on AZ31 alloy. Anti-Corrosion Methods and Materials. 2018;**65**(3):317-324

[42] Montemor MF, Ferreira MGS. Electrochemical study of modified bis-[triethoxysilylpropyl] tetrasulfide silane films applied on the AZ31 Mg alloy. Electrochimica Acta. 2007;**52**(27):7486-7495

[43] Zanotto F, Grassi V, Frignani A, Zucchi F. Protection of the AZ31 magnesium alloy with cerium modified silane coatings. Materials Chemistry and Physics. 2011;**129**(1-2):1-8

[44] Qiao Y, Li W, Wang G, Zhang X. Corrosion protection of AZ31 magnesium alloy treated with La3+ modified

3-methacryloxypropyltrimethoxysilane conversion film. Journal of Rare Earths. 2015;**33**(6):647-654

[45] Figueira RB. Hybrid sol-gel coatings for corrosion mitigation: A critical review. Polymers. 2020;**12**(3):689

[46] Calado LM, Taryba MG, Carmezim MJ, Montemor MF. Self-healing ceria-modified coating for corrosion protection of AZ31 magnesium alloy. Corrosion Science. 2018;**142**:12-21

[47] Gao H, Li Q, Dai Y, Luo F, Zhang HX. High efficiency corrosion inhibitor 8-hydroxyquinoline and its synergistic effect with sodium dodecylbenzenesulphonate on AZ91D magnesium alloy. Corrosion Science. 2010;**52**(5):1603-1609

[48] Toorani M, Aliofkhazraei M, Mahdavian M, Naderi R. Superior corrosion protection and adhesion strength of epoxy coating applied on AZ31 magnesium alloy pre-treated by PEO/Silane with inorganic and organic corrosion inhibitors. Corrosion Science. 2021;**178**:109065

[49] Ashassi-Sorkhabi H, Moradi-Alavian S, Esrafili MD, Kazempour A. Hybrid sol-gel coatings based on silanes-amino acids for corrosion protection of AZ91 magnesium alloy: Electrochemical and DFT insights. Progress in Organic Coating. 2019;**131**:191-202

[50] Tan ALK, Soutar AM, Annergren IF, Liu YN. Multilayer sol-gel coatings for corrosion protection of magnesium. Surface and Coating Technology. 2005;**198**(1-3):478-482

[51] Merino E, Durán A, Castro Y. Integrated corrosion-resistant system for AZ31B Mg alloy via plasma electrolytic oxidation (PEO) and sol-gel processes. International Journal of Applied Glass Science. 2021;**12**(4):519-530

[52] Liu C, Lu X, Li Y, Chen Q, Zhang T, Wang F. Influence of post-treatment process on corrosion and wear properties of PEO coatings on AM50 Mg alloy. Journal of Alloys and Compounds. 2021;**870**:159462

[53] Li N, Zheng Y. Novel magnesium alloys developed for biomedical application: A review. Journal of Materials Science and Technology. 2013;**29**(6):489-502

[54] Staiger MP, Pietak AM, Huadmai J, Dias G. Magnesium and its alloys as orthopedic biomaterials: A review. Biomaterials. 2006;**27**(9):1728-1734

[55] Zhao-Qi Z, Yong-Xin Y, Jing-an L, Rong-Chang Z, Shao-Kang G. Advances in coatings on magnesium alloys for cardiovascular stents—A review. Bioactive Materials. 2021;**6**(12):4729-4757

[56] Durisin M. Bioabsorbable behaviour of magnesium alloys—An in vivo approach. In: Sakara-Narayanan TSN, Song Park I, Min-Ho L, editors. Handbook of Surface Modification of Magnesium and its Alloys for Biomedical Applications. Sawston: Elsevier, Woodhead publishing; 2015. pp. 123-178

[57] Sanchez AHM, Luthringer BJC, Feyerabend F, Willumeit R. Mg and Mg alloys: How comparable are in vitro and in vivo corrosion rates? A review. Acta Biomaterialia. 2015;**13**:16-31

[58] Heimann RB, Lehmann HD. Recent research and patents on controlling corrosion of bioresorbable Mg alloy implants: Towards next generation biomaterials. Recent Patents on Materials Science. 2017;**10**(1):2-19

[59] Gaur S, Singh Raman RK, Khanna AS. In vitro investigation of biodegradable polymeric coating for corrosion resistance of Mg-6Zn-Ca

alloy in simulated body fluid. Materials Science and Engineering: C. 2014;**42**:91-101

[60] Gaur S, Nigam S, Khanna A, Singh Raman RK. Silane-coated magnesium implants with improved in-vitro corrosion resistance and biocompatibility. Journal of Materials Science and Surface Engineering. 2016;**4**(5):415-424

[61] Zomorodian A, Brusciotti F, Fernandes A, Carmezim MJ, Moura T, Fernandes JCS, et al. Anti-corrosion performance of a new silane coating for corrosion protection of AZ31 magnesium alloy in Hank's solution. Surface and Coating Technology. 2012;**206**(21):4368-4375

[62] Castro Y, Durán A. Control of degradation rate of Mg alloys using silica sol–gel coatings for biodegradable implant materials. Journal of Sol-Gel Science and Technology. 2019;**90**(1):198-208

[63] Wang C, Shen J, Zhang X, Duan B, Sang J. In vitro degradation and cytocompatibility of a silane/$Mg(OH)_2$ composite coating on AZ31 alloy by spin coating. Journal of Alloys and Compounds. 2017;**714**:186-193

[64] Dou J, Yu H, Chen C. Preparation and characterization of composite coating on Mg-1.74Zn-0.55Ca alloy by micro-arc oxidation combined with sol-gel method. Materials Letters. 2019;**255**:126578

[65] Gao F, Hu Y, Li G, Liu S, Quan L, Yang Z, et al. Layer-by-layer deposition of bioactive layers on magnesium alloy stent materials to improve corrosion resistance and biocompatibility. Bioactive Materials. 2020;**5**(3):611-623

[66] Liu T, Rui S, Li S. Layer-by-layer self-assembly composite coatings for improved corrosion and wear resistance of mg alloy for biomedical applications. Coatings. 2021;**11**:515

[67] Witecka A, Yamamoto A, Dybiec H, Swieszkowski W. Surface characterization and cytocompatibility evaluation of silanized magnesium alloy AZ91 for biomedical applications. Science and Technology of Advanced Materials. 2012;**13**(6):1878-5514

[68] Amaravathy P, Sowndarya S, Sathyanarayanan S, Rajendran N. Novel sol gel coating of Nb2O5 on magnesium alloy for biomedical applications. Surface and Coating Technology. 2014;**244**:131-141

[69] Sepulveda P, Jones JR, Hench LL. In vitro dissolution of melt-derived 45S5 and sol-gel derived 58S bioactive glasses. Journal of Biomedical Materials Research. 2002;**61**(2):301-311

[70] Omar SA, Ballarre J, Castro Y, Martinez Campos E, Schreiner W, Durán A, et al. 58S and 68S sol-gel glass-like bioactive coatings for enhancing the implant performance of AZ91D magnesium alloy. Surface and Coatings Technology. 2020;**400**:126224

[71] Nikbakht A, Dehghanian C, Parichehr R. Silane coatings modified with hydroxyapatite nanoparticles to enhance the biocompatibility and corrosion resistance of a magnesium alloy. RSC Advances. 2021;**11**(42):26127-26144

[72] Höhn S, Virtanen S, Boccaccini AR. Protein adsorption on magnesium and its alloys: A review. Applied Surface Science. 2019;**464**:212-219

[73] Liu X, Yue Z, Romeo T, Weber J, Scheuermann T, Moulton S, et al. Biofunctionalized anti-corrosive silane coatings for magnesium alloys. Acta Biomaterialia. 2013;**9**(10):8671-8677

[74] Majumder O, Singh Bankoti AK, Kaur T, Thirugnanam A, Mondal AK. The influence of silane and silane–PMMA coatings on the in vitro biodegradation behavior of AE42 magnesium alloy for cardiovascular stent applications. RSC Advances. 2016;**6**(109):107344-107354

[75] Xue K, Liang LX, Cheng SC, Liu HP, Cui LY, Zeng RC, et al. Corrosion resistance, antibacterial activity and drug release of ciprofloxacin-loaded micro-arc oxidation/silane coating on magnesium alloy AZ31. Progress in Organic Coating. 2021;**158**:106357

Chapter 5

Experimental Investigation of Mechanical and Wear Behaviour of AZ91 Magnesium Hybrid Composite Materials

Palanivel Mathiazhagan and S. Jayabharathy

Abstract

In recent years, emerging requisite for advanced materials gave a path for hybrid composites. Magnesium metal matrix composites are gaining more interest and a better substitute for heavier steel, aluminium, titanium and even for plastic based materials. At present the AZ91 magnesium alloy is most widely in transport vehicle industry. However, the application of AZ91 magnesium alloys are limited due to several negative effects such as poor creep resistance, wear resistance and inferior corrosion resistance when it is exposed to atmospheric conditions. Future to improve the strength, better corrosion resistance and wear resistance are important for their extend applications of exciting alloy AZ91. The main objective of the present investigation is to achieve above mentioned properties. The AZ91 alloy was reinforced with titanium dioxide/0.5% graphene and with titanium/0.5% graphene in varying weight percentage (1%, 2%) by stir casting technique. These combinations are called hybrid metal matrix composite of materials such as AZ91 + 1%Ti +0.5% Gr (A1), AZ91 + 2% Ti +0.5% Gr (A2), AZ91 + 1%TiO2 + 0.5% Gr (B1) and AZ91 + 2%TiO2 + 0.5% Gr (B2) alloys. The following experiments such as tensile, compressive, hardness and wear tests have been carried out to find all the properties from the newly developed hybrid metal matrix composite of materials and compared with AZ91. Wear tests have been carried out by pin on disc tribometer for both dry and wet sliding condition under 20 N,40 N,60 N, and 80 N. The results indicated the AZ91–1%TiO2–0.5%Gr having high wear resistance compared to other three combinations as well as AZ91. The present experimental investigations of hybrid metal matrix composite of materials have wear resistance in the order of B1 > A2 > A1 > B2 > AZ91 and AZ91–2%TiO2–0.5% Gr showed good tensile strength and hardness. The enhanced these properties were discussed in this paper.

Keywords: hybrid composites, titanium dioxide, titanium, graphene and Wear

1. Introduction

Materials are the key stuff of engineering to household applications which creates pursuit for commercialization. The present world is a competitive customer driven one. The need to save our mother Earth, limited primary resources and stringent norms paved the way to use of the newer materials in engineering sectors. Quality, comfort, safety and reliability of product are the major concern of survival of the industries. Product's quality is influenced by its design features, material selection and processing techniques. This emphasised the innovation of newer materials to tailor the needs of ever growing demand product for better engineering properties as well as aesthetics. Materials are playing the most vital part from the past till the current scenario as well as in upcoming future. These are substances out of which anything can be prepared. They have contributed to the development of various fields such as space, communication, transportation, medicine, agriculture and food processing industries. Light weight magnesium and its alloys are the most promising next generation materials. Magnesium and its alloys find unique place in the automotive sector due to better solidification and fluidity, high strength to weight ratio, good damping behaviour and high recycling potential in automotive, aerospace, defence and electronics for better radiation shielding effect as well in biomedical sectors as properties of bone is similar to magnesium.

1.1 Classification of materials

Most of the engineering materials are broadly classified into one of the following such as Metals and alloys (ferrous & non-ferrous), Ceramics, Organics, Semiconductors, Biomaterials and Composites.

1.2 Metals and alloys

Metals are the class of materials characterised by their high electrical conductivity as well by thermal conductivity. Metals are crystalline solids which are closely packed and symmetric in nature. Based on the crustal abundance, metals are arranged as - aluminium, iron, calcium, sodium, titanium, copper, brass, magnesium. New materials were developed from predates by intrinsic modification that is by addition of alloying elements to enhance the mechanical and wear properties of the existing materials.

1.3 Ceramic materials

Any inorganic, non-metallic solid substances that has chemical inertness, high strength, hardness, high melting temperature, low thermal and electrical conductivity as their characteristics but brittle in nature may be termed as ceramics. Alumina, calcia are few examples.

1.4 Composite materials

New materials are developed to meet the customer driven competitive environment by extrinsic modification. Composite materials fall under this category. Composite materials constitute two or more distinct type of materials. Reinforcements are

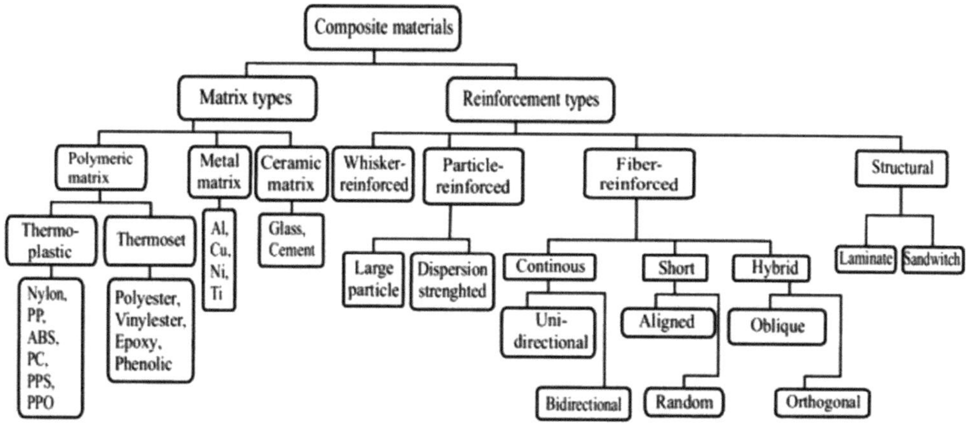

Figure 1.
Classification of composite materials based on reinforcement and matrix types.

added to the parent material to improve the properties. **Figure 1** show the classification of composite materials.

2. Need for composite materials

Pollution has become the major consideration by legislation to save our mother earth, in particular, the emission of carbon dioxide by combustion of fossil fuels. Globally, the transportation industries face two major issues - CO_2 emission and fuel consumption. Moreover, stringent rules framed by the environment pollution act [1] to reduce CO_2 emission and for energy conservation, have paved the way for automotive sectors to turn their attention towards energy efficient vehicles. On the other hand, consumers demand for better aesthetic appearance, comfort, use of electronic system like image processing, navigators and extra safety measure etc., adds weight to the existing system [2].

Some of the possible ways to resolve the issues faced by automobile manufacture sector are: Innovation in use of green energy, Improvising the design and Substitution of advanced material with the existing ones. The use of green energy includes high technology incubation and high cost. Design improvement again involves a lot of research and development from the point of industry. So to meet out the growing demands and challenges, automotive sectors are focusing on use of advanced lightweight materials as a better solution.

Enhancing use of light weight material in current scenario is not only most significant way to reduce the overall weight of the vehicles but also cost effective one. Another important criterion to be considered by reduction in weight of vehicle results in reduction in greenhouse gases emission. It is evident that 100 kilogrammes of self - weight of the vehicles reduced gives us 0.5 litres per 100 kilometres as fuel saving [3].

To achieve weight reduction, the most promising lightweight materials were used to directly substitute heavier steel. Among the various materials, designers found magnesium has potential to substitute heavier steel and aluminium [4]. Gradually magnesium and its alloys usage has increased greatly in the last few decades [5]. It is evident from various researches that magnesium has growing demand in automotive sectors, biomedical, electronics due to its unique characteristics [6]. Another major issue is regarding the radiation by using of mobile phones apart from energy efficiency, stringent norms of pollution control and use of limited resources. Due to new regulation concerning the electromagnetic radiation effect of electronic gadgets paved the way for higher demand for magnesium die cast parts since magnesium has excellent electromagnetic shielding effect [7].

Magnesium and its alloys has low density, high strength to weight ratio, good castability, better solidification due to low latent heat, good vibration damping effect (high speed application), recyclable, better noise dampening, good manufacturability than the conventionally used metals [8]. Among the various magnesium based alloys, AZ (Aluminium and Zinc) and AM magnesium alloy finds wide spread applications.

However, the application of magnesium alloys is limited due to some negative effects like poor creep resistance, and inferior corrosion resistance when exposed to sea and road salt [9].

Composite materials provide an opportunity to enhance particular property along with the advantages of base material. New spot in advanced innovation is hybrid metal matrix composite materials [10] which tailors the need of the functional requirement. Selection of proper material is a vital part of design team in automobile companies. They focus on performance as well as cost. Hybrid composite materials are combination of composite and hybrid materials. They possess two or more reinforcements having different properties. The performance of the hybrid composite material is the joint effect of all the individual constituents. They are usually employed for low density and high strength applications particularly in the field of automotive sector and aerospace sector.

Most of the magnesium alloys possesses excellent fluidity, good machinability and processability which enables the production of complicated die cast parts easier [11]. AZ91(Mg-9Al-0.8Zn-0.2Mn) has less susceptibility to hydrogen porosity and less reactive iron hence can be poured in steel mould. Among the various alloys of magnesium, AZ91 is conferred as one of the best magnesium alloy suitable for thin walled applications. Use of AZ91 instead of aluminium reduced the weight of about 25% but the geometry and production tools remained the same compared to AZ [4]. In addition, literatures also reveal that most of published work were focused on magnesium based composites owing to their good mechanical properties, good damping behaviour which arrests the vibration and reduces the sonic emission and good radiation shielding effect [11]. The SiC, Graphene Nanoplates, Boron carbide and titanium reinforced with Magnesium alloy to improve the mechanical properties.

3. Scenario of composite materials and magnesium alloys

Light weight material still plays a significant role in structural application in the field of automotive and aerospace sectors. Demand for newer materials increases every year which is evident in various scientific researches carried all over world by researchers. It has been possible because of growing technologies, processing techniques, advent of newer phase in materials and its applications in this customer's demand driven world. Composite materials have brought a revolution in the world of material. These materials are unique and different from the monolithic or heterogeneous materials. Composite materials constitute two or more distinct type of material which is not the by-product of chemical reaction between the constituents. It has two phases (constituent material), one of the constituent is reinforcing phase which are discontinuous and are embedded in continuous phase called matrix.

Matrix is continuous phase whose work is to retain the shape of the composite, share the stress to other phase. They provide better finish to the product.

Reinforcement phase is strong and may not be less dense. Reinforcing phase may be in the form of fibres, particles or flakes. They contribute the desired properties and also transfer the strength to the matrix.

3.1 Types of composites

The composites materials are classified into three major types.

Polymer matrix composites.

Ceramic matrix composites.

Metal matrix composites.

3.2 Polymer matrix composites

Polymer Matrix Composites (PMCs) has organic polymer as matrix and reinforcing phase are a variety of short or continuous fibres. Organic polymers are long chain of carbon. Fibres are used to enhance the mechanical properties and to carry loads. Matrix's purpose is to bond the fibres and protect reinforcing phase. Polymer matrix composites are categorised into two types based on mechanical properties (strength and stiffness): reinforced plastics and advanced composites. Reinforced plastics are polymer matrix composites that impart additional strength by adding embedded fibrous material into plastics. These type of composites are usually cheap and consist of polyester resins reinforced with glass fibres (low stiffness). Advanced composites consist of matrix and fibres which facilitates superior stiffness and strength. Fibres in advanced composite type are mostly high performance like graphite, aramid. Advanced composites are relatively expensive but possesses high strength and superior stiffness.

3.3 Ceramic matrix composites

Ceramics are solid materials which exhibit very strong ionic bonding and in few cases covalent bonding. Ceramic matrices such as silicon nitride, silicon carbide surround the fibres. They are the good choice for high temperature applications. Ceramic matrix composites possess principal characteristics like resilience to oxidation and high compressive strength. Production errors or scratches may lead to cracks in conventional ceramics. However in ceramic matrix composite, by embedding the fibres, fracture toughness values is enhanced. Even crack propagation was hindered by the presence of fibres (reinforcement) in matrix of ceramic composites.

3.4 Metal matrix composites

Metals play significant and versatile role extending itself from house hold to engineering application. Metal Matrix Composite (MMC) as the name implies has reinforcement (fibres or particles) surrounded by matrix of metal. The matrix is the monolithic material. The main matrix materials used are aluminium, titanium, magnesium and copper. Reinforcement phase not only carries the structural task but also used to enhance the property like wear resistance, resistance to degradation by fluid and so on. Metal matrix composite finds its wide spread opportunities as it tailors the need to achieve a better property beyond those unattainable by monolithic materials. These composites potentially have high stiffness, strength and high temperature resistance over polymer matrix composites. However metal matrix composites can operate in varied temperatures and has better electrical and thermal conductivity.

In recent years, metal matrix composite materials are gaining more attention due to their greater wear resistance, higher thermal conductivity when compared to unreinforced metals and alloy. Apart from these advantages, Metal matrix composites are environmentally suitable as they do not absorb water like PMC. They use low density materials like aluminium and magnesium as matrix.

3.5 Hybrid composites

Hybrid composite materials are consisting of a matrix and possess two or more reinforcements having different properties. The performance of the hybrid composite material is the joint effect of all the individual constituents. They are usually employed for low density and high strength applications particularly in the field of automotive sector and aerospace sector.

3.6 Magnesium based composites

Magnesium alloys and its composites find a significant place as next generation materials as a substitute for aluminium and steel. Magnesium offer better mechanical and physical properties like high strength, good vibration dampening capacity, better castability, ease of machinability, good recyclability, environmental friendly and above all low density 1.74 g/cc (pure magnesium) than the conventionally used materials in the automotive sectors like iron and aluminium [2]. Pure Magnesium though light in weight but has limited use due to its insufficient strength and corrosion resistance [12]. Alloying magnesium with one or more element like aluminium, zinc, manganese, zirconium, rare earths and so on gave rise to new phase of magnesium alloys which enhanced the properties of pure magnesium. Magnesium alloys not only stand for their better properties but also as environmental friendly material [13].

Among various grades of magnesium alloys, A-Z magnesium alloy are used for structural applications. It has good castability and high specific strength. AZ91 magnesium alloy finds numerous applications in building structural components in aerospace and automotive sector like chassis, wheel, and indoor frame so on [14]. But its negative effect is poor corrosion resistance, low wear resistance in some service environments. It exhibits poor corrosion resistance property in sea salt environment but generally has good corrosion resistance during atmospheric exposure condition [15]. However, its application cannot extend beyond the limit due to some of its negative effects. Reinforcement of AZ91 magnesium alloy with suitable particles can overcome the shortfalls of alloy. Due to growing demand and to meet out the requirements, hybrid composite material is more advantageous than the conventional and composites materials. There are various researches on magnesium metal matrix composite reinforced with SiC, Al_2O_3, Ti, TiO_2, Carbon nanotubes, Graphene, TiC to establish the correlation between mechanical properties and microstructure characteristics of composites, corrosion resistance, wear resistance property [16] etc. It was observed that there were not much research reports on the influence of titanium oxide and graphene and titanium & graphene reinforced with AZ91 magnesium alloy as hybrid composites.

3.7 Selection of matrix material: AZ91 magnesium alloy (Mg-9Al-0.8Zn-0.2Mn)

Constituents of composite materials are matrix and reinforcement. Matrix plays a major role in binding and protecting the reinforcement, distributing the load, enhancing the final property and giving better finish to final product. Matrix is usually aluminium, magnesium and titanium for structural applications.

AZ91 magnesium alloy possesses good mechanical properties. It has less susceptibility to hydrogen porosity and less reactive iron that can be poured in steel mould. Among the various AZ magnesium alloy, AZ91 is conferred as one of the best magnesium alloy suitable for thin walled applications. Usage of AZ91 magnesium alloy instead of aluminium alloy reduced the weight by about 25%, but the geometry and production tools remained the identical [17].

3.8 Reinforcement

Reinforcement can enhance desired properties of the conventional materials. The desired properties depend upon the working environment in which the material is exposed like strength, stiffness, corrosion resistance, wear resistance, reduced weight, dimensional stability, fatigue and extends up to attractiveness and cost. Reinforcements can be classified into two groups: continuous and discontinues. Metal matrix Composites processed by them are termed as continuously reinforced metal matrix composite and discontinuously reinforced composite. They are further sub divided into categories like short fibres, whisker, particulates, continuous fibre and preform/wire (fibre solid). Some important parameters considered are as follows for the selection of reinforcement [15].

- Geometry-Size
- Shape –spherical, flake like, needle
- Content –Volume
- Poly or single crystal
- Inherent properties – Strength, density, hardness

In order to attain the desired properties in (AZ91 magnesium alloy) like improved wear resistance and corrosion resistance, right choice depends upon the types of reinforcement, production method and compatibility with matrix. Most widely used reinforcement in Metal matrix composites is ceramics [18]. Usually ceramics such borides, oxides and carbides are widely used reinforcements.

One of the best advantage of Magnesium/Magnesium alloys based composites over aluminium composites is that the wettability of liquid magnesium with ceramics reinforcement phase is better. Use of titanium reinforcement in Magnesium MMCs gave better results which had the potential to create a new material with excellent mechanical properties, wear and corrosion resistance [19]. Carbon nanotube and graphene are also gaining high interest in the field of materials. Graphene possesses better mechanical, thermal and electrical properties and also light weight with density of 1.06 g/cc. Addition of graphene as reinforcement in magnesium composites has enhanced the strength of metals. Titanium oxide are usually used for coating purpose to enhance corrosion resistance property. Titanium oxide used as reinforcement with magnesium matrix showed improved hardness and strength [20].

3.9 Selection of reinforcement

Selection of reinforcement plays a crucial role as improving the desired properties of final material. As reinforcement (discontinuous or continuous) is embedded in the matrix, they share the load and transfer strength to the matrix. Discontinues metal matrix composites are mostly isotropic in nature and also cost effective. The following reinforcements are selected for this work.

Graphene.
Titanium and
Titanium dioxide.

Most widely used and recommended reinforcement were titanium carbide, boron carbide, MWCNT for melting metallurgical methods. Yet the studies discloses the other types of reinforcement like titanium dioxide and graphene are usually used for coating purpose to enhance the corrosion resistance property on magnesium alloy and results obtained were positive. But titanium dioxide and graphene used as reinforcement in magnesium matrix hybrid composites are limited and yet to be explored. Similarly, titanium and graphene as reinforcements embedded in magnesium matrix are yet to be explored.

4. Wear

Tribology is the branch of science dealing with contacting surfaces in relative motion. It deals with friction, lubrication and wear of interacting surfaces. Wear is recognised as the phenomenon of undesirable removal of materials from the surface due to interaction with the counter surface. Wear resistance is not the basic property of any material, but depends on environment. It varies even if it is not exposed to the same work environment. Almost all moving parts of machines durability, reliability and life depend upon one main parameter: wear. Therefore, wear property demands its analysis as strong need for advanced and reliable material property for application oriented environment. Wear can be controlled by the selection of right material and proper operating conditions [21]. Wear is a critical factor composing of dynamic parameters, environmental parameters, and material parameters. So wear resistance property is usually analysed for purpose of lubrication, to screen material & its surface treatments, sometime to establish relationship between finishing methods (processing) and its performance. The different type of wear is sliding wear, impact wear and rolling contact wear. Sliding wear is a relative movement between two solid surfaces acting tangential to each other.

5. Materials

The AZ91 magnesium alloy is the base material. AZ91 magnesium alloy used in the present work was fabricated by stir casting process. This alloy was prepared by commercial magnesium ingots, aluminium ingot and zinc ingot. AZ91 magnesium alloy has main composition of 9% Aluminium, 1% Zinc and 0.3% Manganese. Aluminium of 9% by weight is added to enhance the castability, hardness and as well the freezing range. Zinc is added upto 1% by weight to magnesium as alloy to increase harmful corrosive effects. Zinc when exceed more than 1% will cause hot shots. Manganese (0.3% by weight) was added to increases the resistance against corrosion.

Element	Al	Mn	Zn	Si	Fe	Others	Mg
Composition	8.3–9.1%	0.15%	0.35–1.0%	0.01%	<0.005%	<0.052%	Balance

Table 1.
Nominal composition of AZ91 magnesium alloy.

The remaining was pure magnesium. Chemical composition of AZ91 magnesium alloy is listed in **Table 1**. Titanium, titanium dioxide and graphene were the reinforcements used for the preparation of hybrid composite materials. Titanium (99.9% supplied by Nextgen steel and suppliers, Mumbai) graphene nanoparticles of 99% purity (supplied by AdNano Technologies, Karnataka) and titanium dioxide of 250 nm (99.97% purity supplied by Sisco Research laboratory (SRL) Maharashtra). The AZ91 magnesium alloy and hybrid composites are manufactured by stir casting technique.

5.1 Fabrication of base material AZ91 magnesium alloy (matrix)

Initially the ingots are cut and weighed for right proportion. The proportions are illustrated in **Table 2**.

Material	% weight	Weight
Magnesium	89.7%	897 grams
Aluminium	9%	90 grams
Zinc	1%	10 grams
Manganese	0.3%	3 grams

Table 2.
Proportion of base material (matrix) AZ91.

5.2 Hybrid composites by stir casting process

A recent development in stir casting is a two-step process. The effectiveness of the two-step method of processing is mainly attributed to its ability to break the gas layer around the particle surface. Particles have a thin layer of gas being absorbed on their surface, which hinders wetting between the particles and molten metals.

Magnesium alloys has unique feature in solidification. In this process, the metal is heated to above its liquids temperature with inert argon gas atmosphere, because Magnesium is flammable in molten state. Then the metal is fully melted. The melt is then cooled to a temperature between the liquids and solidus points. It is now in a semi-solid state. At this stage, the preheated particles are added and stirred well by mechanical stirrer for certain period of time to ensure proper mixing. The slurry is again heated and mixed thoroughly. This technology is relatively simple, easy adaptable and cost effective.

The test samples were cut into desired size (10 x10x 5) mm and polished with silicon carbide paper up to 1200 grit. Alumina powder polishing was done to ensure mirror image on the specimens. The prepared specimens are washed with acetone and dried. The specimens are then etched with acetic picral (5 mL acetic acid, 6 g picric acid, 10 mL water, 100 mL ethanol (95%) a universal etchant). The test sample was immersed gently until the face turns brown. It was then washed with ethanol and air dried at room temperature.

6. Mechanical testing

The utility of metals/composites in different fields of engineering depends on its capacity to accomplish design and service requirements of the current scenario. The competence of composites to satisfy these requirements is determined by the mechanical, thermal and physical properties of the metal [22]. Mechanical testing like tensile test, compression test and hardness test were done on the hybrid composites with varying composition and varying weight % of reinforcements as shown in **Table 3**.

Sr.No	Hybrid composite Sample	AZ91 (%)	Ti (%)	TiO2 (%)	Gr (%)
1	AZ91 + 1%Ti +0.5% Gr (A1)	98.5	1	—	0.5
2	AZ91 + 2%Ti +0.5% Gr (A2)	97.5	2	—	0.5
3	AZ91 + 1%TiO2 + 0.5% Gr (B1)	98.5	—	1	0.5
4	AZ91 + 2%TiO2 + 0.5% Gr (B2)	97.5	—	2	0.5

Table 3.
Proportions of matrix and reinforcements for hybrid composites.

6.1 Tensile test

The tension test is one of the most common tests performed on the casted samples for evaluating materials. The tension test was done by gripping opposite ends of a tensile test sample within the load frame of a tension machine. A tensile force was applied by the machine, resulting in the gradual elongation and then results in fracture of the test sample. The mechanical properties found from tension test are yield strength, ultimate tensile strength, ductility properties, such as elongation and reduction in area. **Figure 2** shows the dimension of test specimen as per ASTM and **Figure 3** shows the tensile test sample of hybrid Composite specimen.

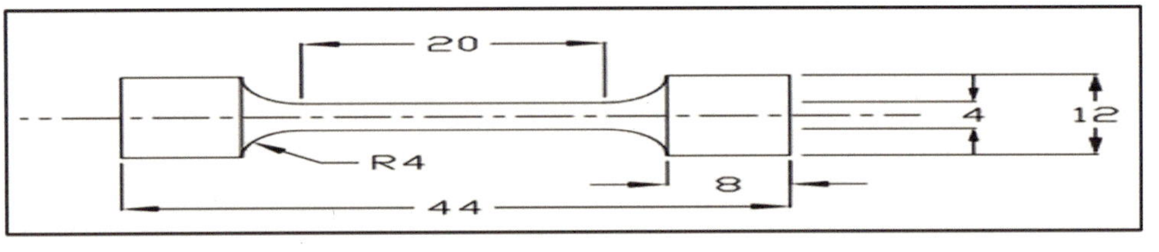

Figure 2.
Dimensions of test specimen-ASTM.

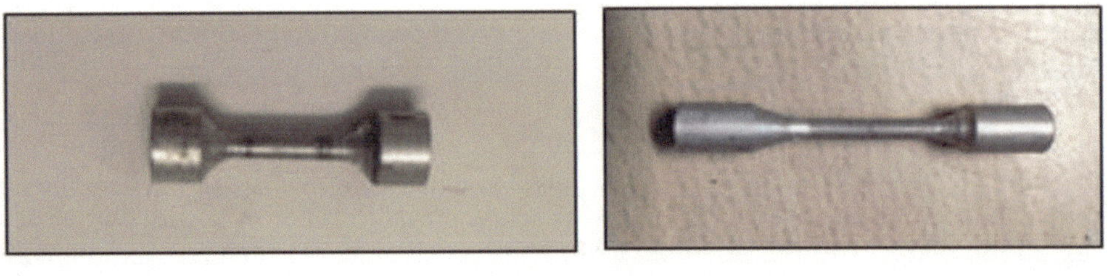

Figure 3.
Tensile test sample of hybrid composite prepared as per standards.

7. Results and discussion

7.1 Microstructural characterisation

Microstructural studies conducted on the casted hybrid composites sample revealed the equiaxial grain morphology and monolithic showed coarse grain structure. The unreinforced AZ91 showed coarse dendrites and SEM observation indicated the presence of the α-Mg matrix which was surrounded by inter dendrite and β phase ($Mg_{17}Al_{12}$). Further SEM image with EDS analysis was used to verify the micro-composition of the unreinforced AZ91 magnesium alloy. The EDS analysis shows the presence of magnesium, aluminium and zinc peaks but manganese cannot be detected due to low content of addition (0.3%) as shown in **Figure 4**.

Addition of titanium and graphene to AZ91 magnesium has positively refined the grain size. The AZ91 magnesium alloy, AZ91 + 1% Ti + 0.5%Gr and AZ91 + 2% Ti + 0.5%Gr showed the intermetallic compounds appeared brighter surrounding the grains. The titanium combines with aluminium during initial stage of solidification. Hence at eutectic temperature there is less content of aluminium available to form β phase ($Mg_{17}Al_{12}$) [23]. The EDS analysis proves the presence of graphene of the

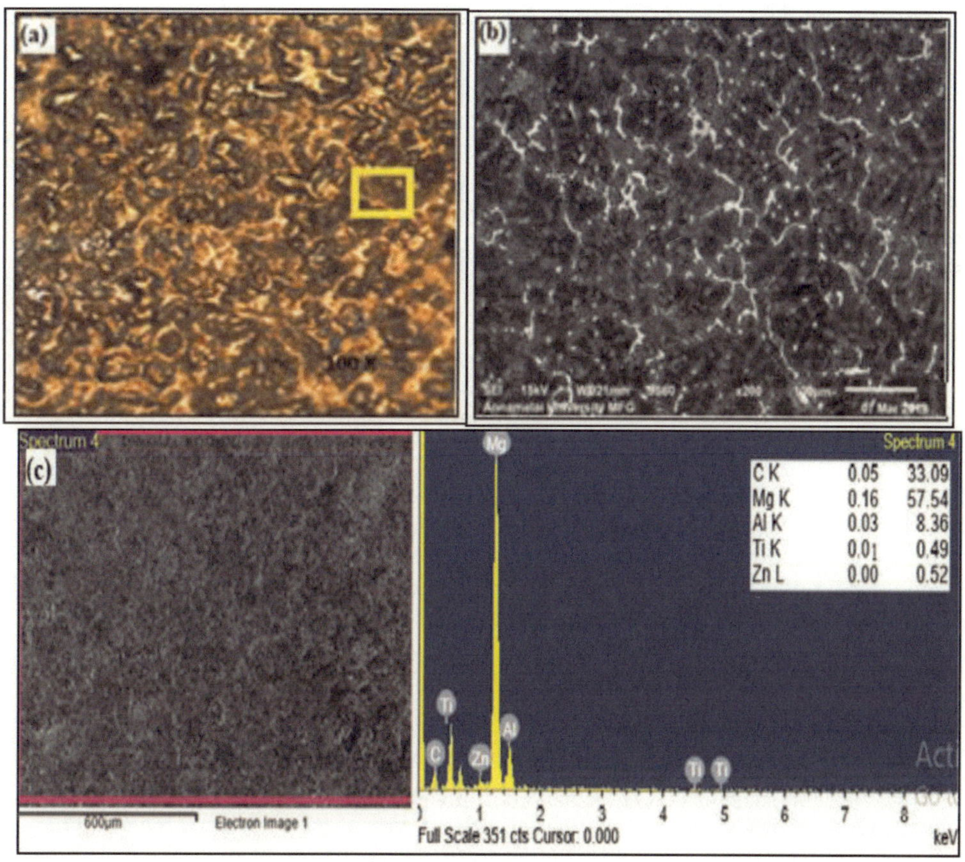

Figure 4.
AZ91/1% Ti/0.5% Gr (a) optical micrograph (b) SEM image (c) EDS spectrum.

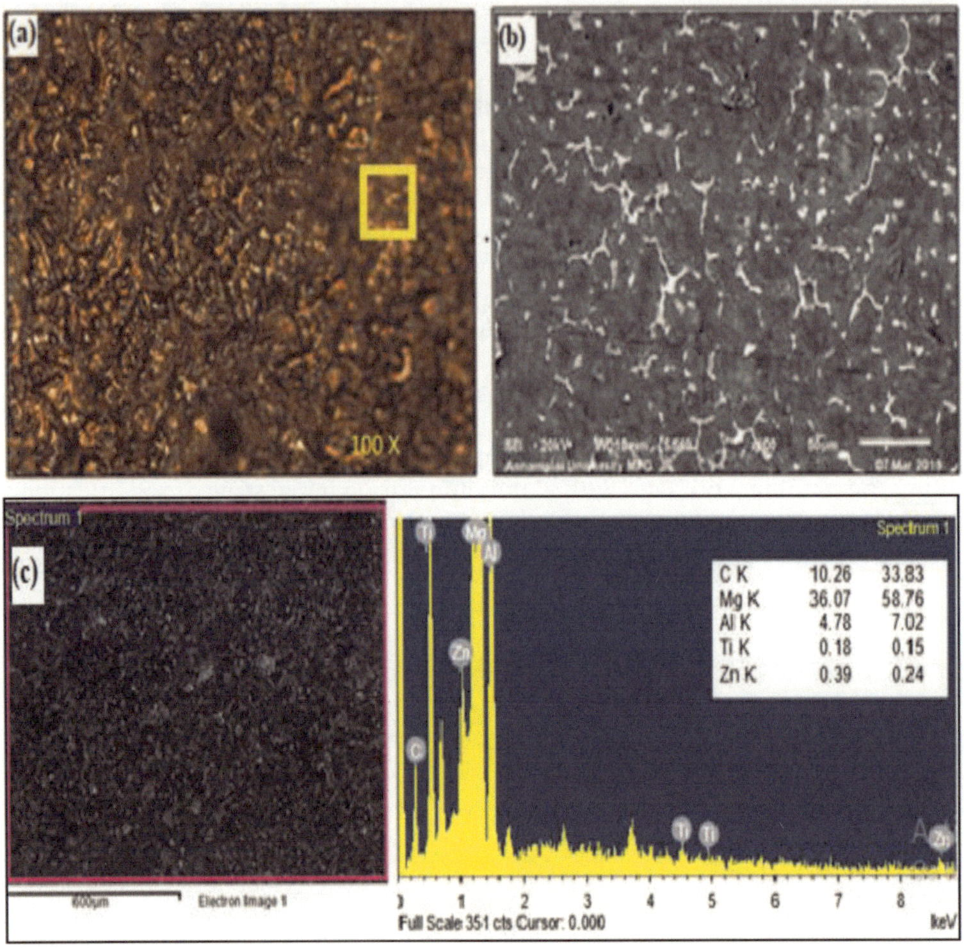

Figure 5.
AZ91/2% Ti/0.5% Gr (a) optical micrograph (b) SEM image (c) EDS spectrum.

hybrid composites [24] depicted in **Figure 4**. The other combination of hybrid composite AZ91 magnesium alloy was with titanium dioxide and graphene also shows the discontinuous β precipitates are found around the grains. The EDS spectrum clearly signifies in the **Figures 4–7** and reveals the presence of magnesium, aluminium, zinc, graphene and titanium dioxide. The titanium dioxide as overlapped with Ti and oxygen. It was observed that the reinforcements are uniformly distributed in the AZ91 magnesium matrix for all the hybrid composites. It was also observed no voids or cracks from the micrographs which is assisted by density measurement.

7.2 Density measurement

The results of density test are illustrated in the **Table 4**. The results showed that denser hybrid composites produced among them was AZ91 + 2%Ti +0.5% Gr. The theoretical density was found to be higher than actual density of the hybrid composites.

It was noted from the result that the densities of each hybrid composites were higher than the base matrix clearly signifying that porosity will be reduced by increase

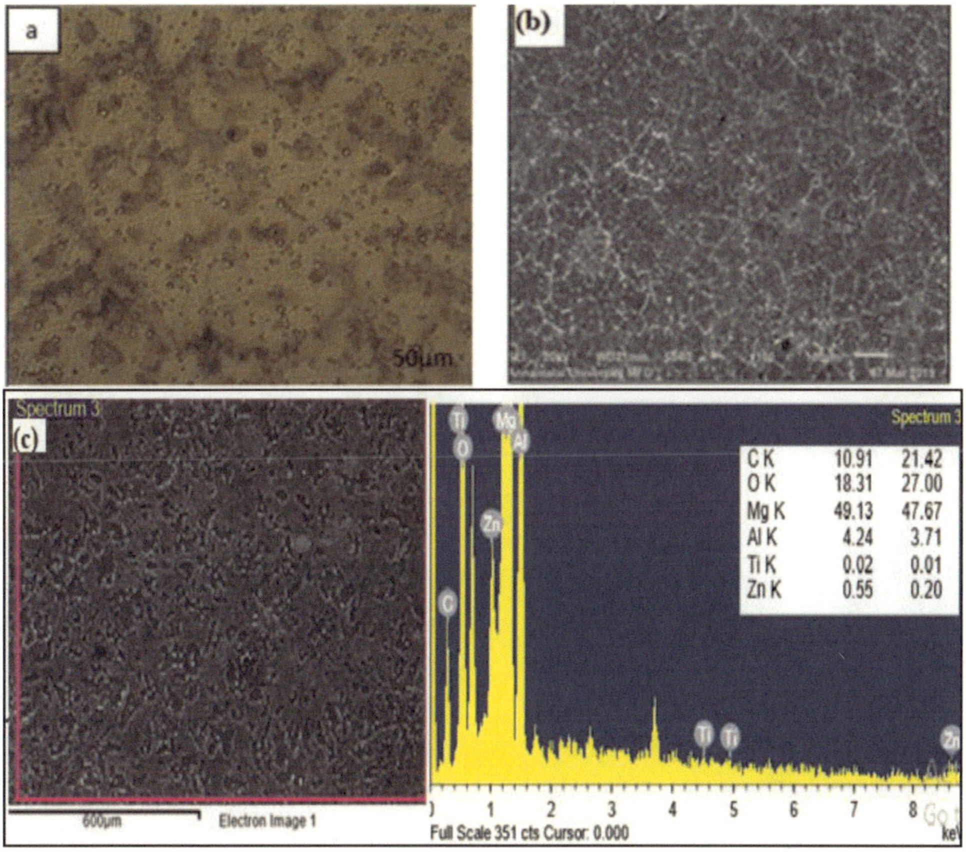

Figure 6.
AZ91/1% TiO_2/0.5% Gr (a) optical micrograph (b) SEM image (c).

the weight percentage of the reinforcement as shown in **Figures 8** and **9**. The increase in density attributes to increase in hardness and decrease in porosity.

7.3 Tensile test

The tensile tests are carried out at room temperature and results of all the hybrid composites are depicted in **Table 5** and **Figure 9**. It revealed that the addition of reinforcement titanium and graphene have gradually increased the strength of the material than the monolithic AZ91. The addition of titanium shows improved ductility than the base matrix. The diffused magnesium and aluminium with titanium yielded higher tensile strength due to the addition of reinforcement graphene, inspite of its high thermal conductivity yet has no chemical reaction with the matrix [25].

The titanium dioxide and graphene hybrid composites exhibited higher strength but probably low ductility. It was noticed that the decrease in ductility may be due to the increase in weight percentage of the reinforcement creating a strained lattice and also void nucleation. The strength of hybrid composites depended on the bonding between the AZ91 magnesium matrix and reinforcements. In both the cases, the graphene acts diffusion barrier for aluminium and magnesium forming intermetallic bond which have poor stability.

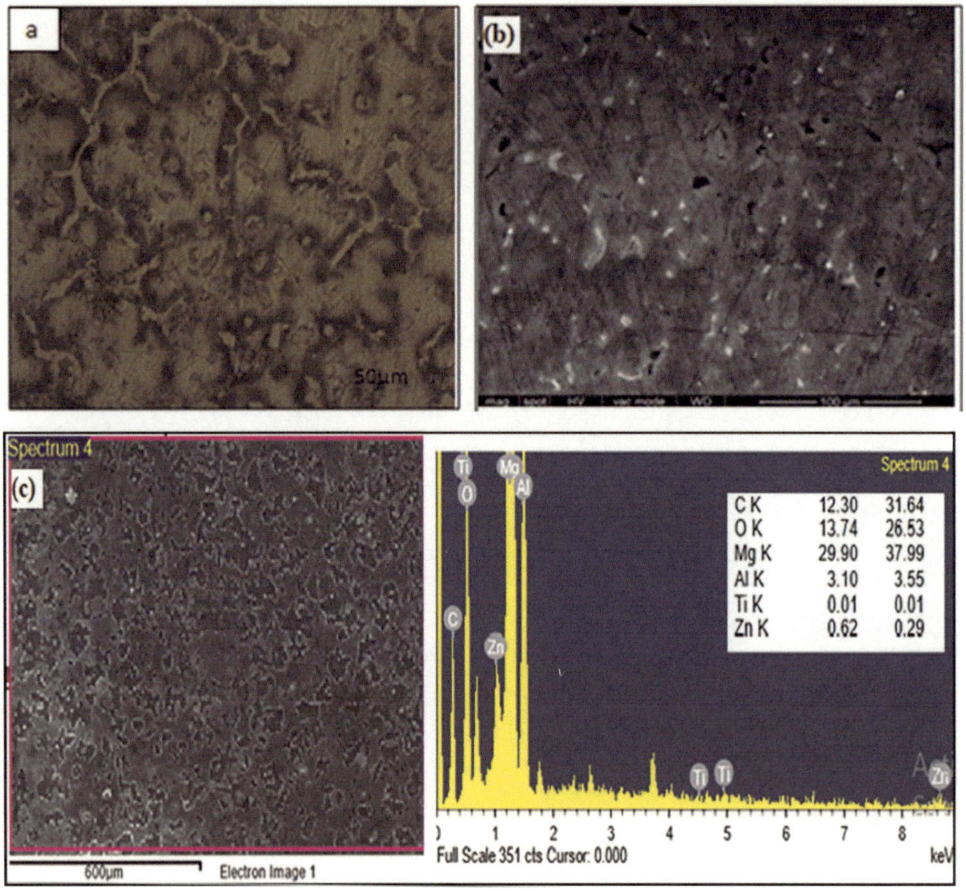

Figure 7.
AZ91/2% TiO_2/0.5% Gr (a) optical micrograph (b) SEM image (c) EDS spectrum.

Sr. No.	Test Samples	Actual density (g/cc)	Theoretical density (g/cc)
1	AZ91	1.81	1.813
2	AZ91 + 1%Ti +0.5% Gr (A1)	1.824	1.826
3	AZ91 + 2%Ti +0.5% Gr (A2)	1.846	1.847
4	AZ91 + 1%TiO2 + 0.5% Gr (B1)	1.823	1.8237
5	AZ91 + 2% TiO2 + 0.5% Gr (B2)	1.841	1.843

Table 4.
Density of hybrid composite material.

7.4 Compression test

Compression test have been conducted at room temperature. The results indicated that significant improvement in ultimate compressive strength of AZ91 magnesium hybrid composite than the unreinforced AZ91 magnesium alloy. Within the hybrid composites sample, the increase in compressive properties was only marginal due to the addition of reinforcements. The test samples and recorded test results are,

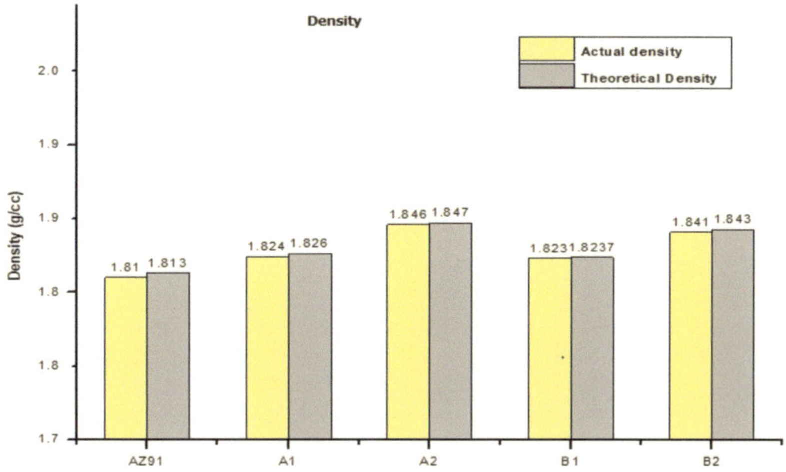

Figure 8.
Density measurement.

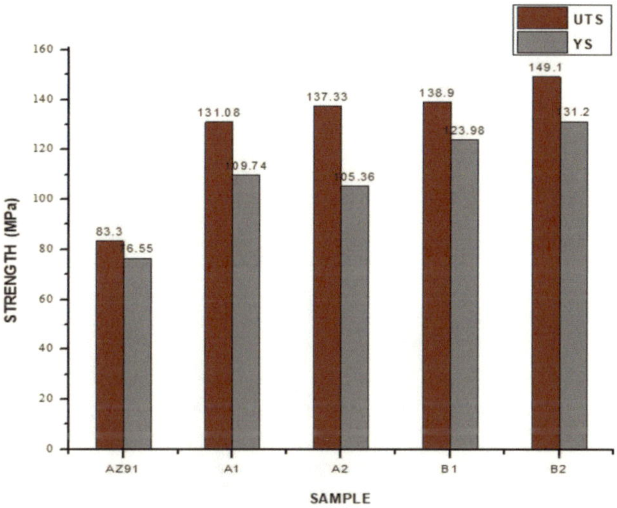

Figure 9.
Tensile strength of AZ91 magnesium alloy and hybrid composites.

illustrated in **Table 6** and **Figure 10**. It was clearly evident that from the compressive testing that hybrid composites and monolithic AZ91 magnesium alloy test samples are split into two halves at angle of 45°. It was observed that reinforcement's combination titanium and graphene with AZ91 have higher compressive strength when compared to the other hybrid combination.

7.5 Hardness

Table 7 shows the test result of micro hardness of hybrid composites and monolithic AZ91 magnesium alloy.

Sr. No.	Test Sample	Ultimate Tensile Strength(MPa)	Yield strength (MPa)	% Elongation
1	AZ91	83.3	76.55	3.0
2	AZ91 + 1%Ti +0.5% Gr (A1)	131.08	109.74	4.52
3	AZ91 + 2%Ti +0.5% Gr (A2)	137.33	105.36	4.56
4	AZ91 + 1%TiO2 + 0.5% Gr (B1)	138.9	123.98	3.5
5	AZ91 + 2% TiO2 + 0.5% Gr (B2)	149.1	131.2	2.94

Table 5.
Tensile strength of AZ91 magnesium hybrid composite material.

Sr. No.	Test Sample	Ultimate Compressive Strength (MPa)	Elongation %
1	AZ91	289.54	2.88
2	AZ91 + 1%Ti +0.5% Gr (A1)	349.9	3.7
3	AZ91 + 2%Ti +0.5% Gr (A2)	366.1	3.29
4	AZ91 + 1%TiO_2 + 0.5% Gr (B1)	335.56	3.4
5	AZ91 + 2% TiO_2 + 0.5% Gr (B2)	351.46	3.36

Table 6.
Compressive strength of hybrid composites.

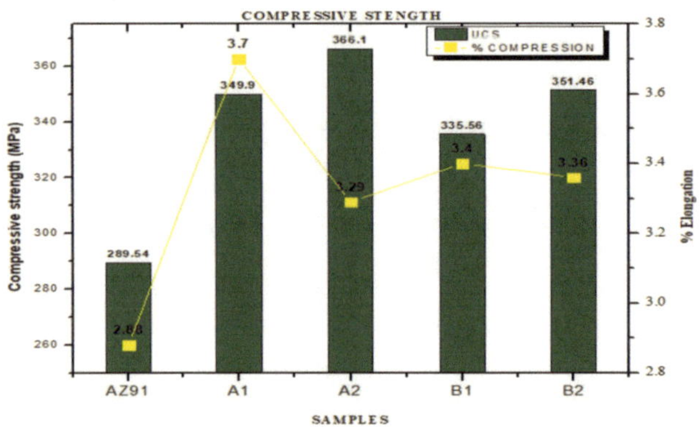

Figure 10.
Ultimate compressive strength.

The results revealed that increase in micro hardness of the reference alloy AZ91 magnesium had increased due to the addition of reinforcements of titanium/graphene and titanium dioxide/graphene. Among the two reinforcement's combinations, it was evident that the hybrid composites titanium/graphene combination exhibited better hardness than the other combination of reinforcement.

The increase in hardness of hybrid composites was due to presence of harder titanium and graphene particles in the reference which assist in load bearing capacity

Sr. No.	Test Samples	Location(1 HV)				
		1	2	3	4	Average
1	AZ91	62.4	62.7	64	66.7	64.0
2	AZ91 + 1%Ti +0.5% Gr (A1)	75	72.3	68.4	74.1	72.5
3	AZ91 + 2%Ti +0.5% Gr (A2)	73.8	80.1	80.7	80.8	78.9
4	AZ91 + 1%TiO_2 + 0.5% Gr (B1)	73.5	71.0	75.2	69	72.2
5	AZ91 + 2% TiO_2 + 0.5% Gr (B2)	70.1	72.2	75.7	75.2	73.3

Table 7.
Hardness.

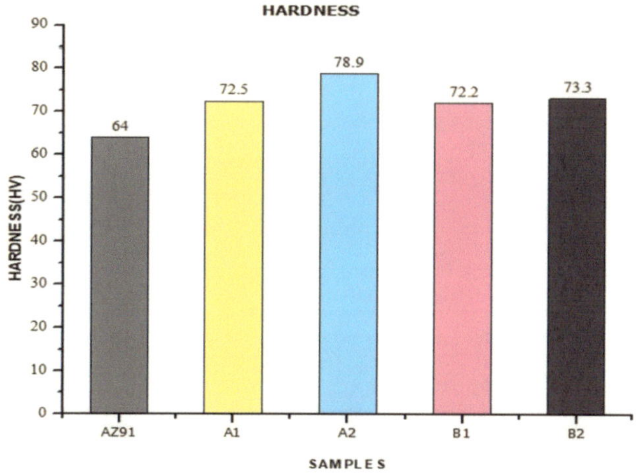

Figure 11.
Hardness.

of matrix [26]. Addition of graphene to AZ91 significantly improved the hardness of the materials. Within the hybrid composites samples, increase in hardness was very significant as the attribute to uniform distribution of reinforcements. The AZ91/2% Ti/0.5% Gr hybrid composite showed the highest hardness and compressive strength among the entire hybrid composite. **Figure 11** shows the micro hardness of all the hybrid composite samples and AZ91 reference alloy.

8. Summary

From the experimental investigation the following results were obtained.

- The yield strength, ultimate tensile strength and compressive strength have significantly increased compared with base alloy with addition of titanium in the range of 1% -2%. The compressive strength has increased by addition of Ti and graphene nano particulates.

Hardness of the entire hybrid composites were increased which attributed to uniform distribution of the reinforcements and load bearing capacity. Titanium and graphene has higher hardness than titanium dioxide and graphene.

9. Wear

9.1 Weight loss

The pin on disc was proposed to study and evaluate the tribological behaviour. The amount of wear loss due to dry sliding and wet sliding conditions of the novel hybrid composites fabricated by using stir casting technique and conducted the experiments. The results of the weight loss of wear test of all the hybrid composites and base alloy AZ91 magnesium are illustrated in the **Tables 8–11**. W1–initial weight of the sample before test(g) and W2-final weight of the sample after test(g). The test results obtained from the dry sliding wear test and wet sliding wear test were thoroughly

Sample	LOAD (N)	WET TEST			DRY TEST		
		W1	W2	Weight loss (g)	W1	W2	Weight loss (g)
AZ91	20	3.2308	3.22	0.0108	2.0953	2.0733	0.022
A1	20	3.8759	3.8699	0.006	3.1271	3.1061	0.021
A2	20	3.366	3.3629	0.0031	1.3662	1.3475	0.0187
B1	20	3.1879	3.1867	0.0012	2.226	2.206	0.02
B2	20	4.2149	4.2115	0.0034	2.512	2.5007	0.0113

Table 8.
The weight loss of the hybrid composites and AZ91 alloy (dry and wet test) with load of 20 N.

Sample	LOAD (N)	WET TEST			DRY TEST		
		W1	W2	Weight loss (g)	W1	W2	Weight loss (g)
AZ91	40	3.6082	3.596	0.0122	2.0733	2.0412	0.0321
A1	40	3.3945	3.3872	0.0073	3.1061	3.0812	0.0249
A2	40	3.603	3.5864	0.011	1.4202	1.3908	0.0294
B1	40	3.5085	3.5054	0.0031	2.2032	2.1782	0.025
B2	40	4.098	4.0873	0.0107	2.5007	2.4727	0.028

Table 9.
The weight loss of the hybrid composites and AZ91 alloy (dry &wet Test) with load of 40 N.

Sample	LOAD (N)	WET TEST			DRY TEST		
		W1	W2	Weight loss (g)	W1	W2	Weight loss (g)
AZ91	60	3.2304	3.2139	0.0165	2.1139	2.077	0.0369
A1	60	3.8701	3.8574	0.0127	2.8955	2.8645	0.031
A2	60	3.3629	3.3461	0.016	1.3669	1.3352	0.0317
B1	60	3.1876	3.1764	0.0112	2.2065	2.1777	0.0288
B2	60	4.2115	4.1985	0.013	2.549	2.517	0.032

Table 10.
The weight loss of the hybrid composites and AZ91 alloy (dry &wet Test) with load of 60 N.

Sample	LOAD (N)	WET TEST			DRY TEST		
		W1	W2	Weight loss (g)	W1	W2	Weight loss (g)
AZ91	80	3.6027	3.5662	0.0365	2.077	2.0374	0.0396
A1	80	3.3927	3.3731	0.0196	2.8645	2.834	0.0305
A2	80	3.5844	3.5601	0.0243	1.3475	1.3175	0.03
B1	80	3.5074	3.4834	0.024	2.177	2.143	0.034
B2	80	4.0783	4.06	0.0183	2.517	2.486	0.031

Table 11.
The weight loss of the hybrid composites and AZ91 alloy (dry &wet Test) with load of 80 N.

analysed to assess the effect of reinforcements and effect of load on the test samples. The sliding distance and velocity were maintained constant. The coefficient of friction is the ratio of frictional force to normal applied load. The coefficient of friction purely depends upon the materials properties and also on surface roughness, temperature of exposure, normal load and environment.

The wear test results (dry sliding & wet sliding) depicted in **Figure 12** (a)-(d) shows the weight loss of base alloy and all the hybrid composites fabricated by stir

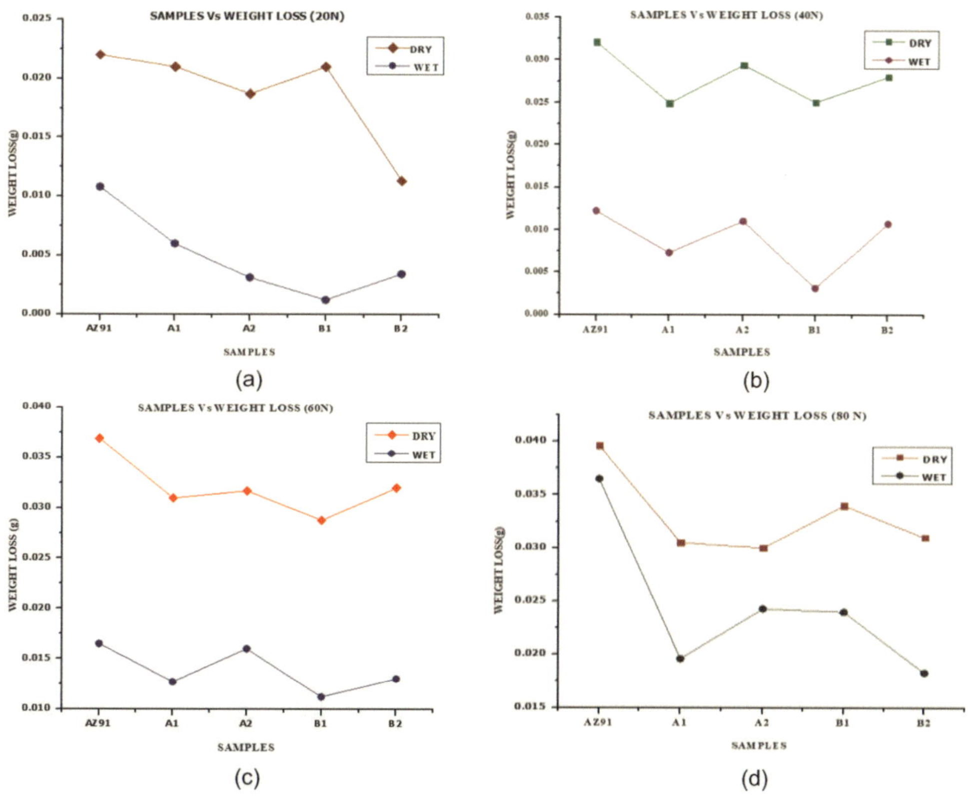

Figure 12.
(a). Samples vs weight loss of hybrid composites and AZ91 (20 N). (b). Samples vs weight loss of hybrid composites and AZ91 (40 N). (c). Samples vs weight loss of hybrid composites and AZ91 (60 N). (d). Samples vs weight loss of hybrid composites and AZ91 (80 N).

casting technique. Hybrid composites exhibited less weight loss under the load condition in both the dry and wet test condition than the reference alloy AZ91.

The lower weight loss may be attributed to the addition of reinforcements (Ti and Gr and TiO_2 and Gr). Among the reinforcement's combinations, TiO_2 and graphene reinforced hybrid composites showed less weight loss than Ti and graphene hybrid composite.

It was also inferred that as normal load increases, the weight loss also increased. Within the hybrid composites the weight loss slightly decreased with increase amount of titanium dioxide (2%). However, titanium and graphene hybrid system, it was found that weight loss increased with increase in the amount of titanium. It was also inferred from the results, weight loss in any test condition of the hybrid composites was not as high as AZ91 magnesium alloy. The similar trend of weight loss was found in dry and wet sliding condition of each hybrid composites.

9.2 Wear rate

The wear rate was calculated for the hybrid composite samples and AZ91 magnesium alloy samples in both dry and wet sliding test conditions under the varying load condition of 20 N, 40 N, 60 N, and 80 N. The results are tabulated in the **Tables 12–15**. The AZ91 magnesium alloy and hybrid composites wear test results of the samples under load 20 N,40 N,60 N, and 80 N in both dry and wet condition revealed that the wear rate, wear mechanism, coefficient of friction and wear resistance of AZ91, A1, A2, B1 and B2 hybrid composite materials.

From the experimental test observation, it was inferred the wear rate was lower for all the hybrid composites than AZ91 magnesium alloy. Also the wear rate was lower for wet sliding than dry sliding condition due to the lubrication effect. The wear rate

Sample	Load (N)	Wet/Wear rate (mm^3/m) x10^{-3}	Dry/Wear rate (mm^3/m)x10^{-3}
AZ91	20	5.295	10.786
A1	20	2.921	10.123
A2	20	1.492	9.500
B1	20	0.585	9.748
B2	20	1.640	7.450

Table 12.
Wear rate of hybrid composites and AZ91 alloy (dry and wet 20 N).

Sample	Load (N)	Wet/Wear rate (mm^3/m)x10^{-3}	Dry/Wear rate (mm^3/m)x10^{-3}
AZ91	40	5.981	15.738
A1	40	3.554	12.121
A2	40	5.989	10.251
B1	40	1.511	12.185
B2	40	5.161	13.505

Table 13.
Wear rate of hybrid composites and AZ91 alloy (dry and wet 40 N).

Sample	Load (N)	Wet/Wear rate $(mm^3/m)x10^{-3}$	Dry/Wear rate $(mm^3/m)x10^{-3}$
AZ91	60	8.090	18.092
A1	60	6.182	14.59
A2	60	7.085	15.256
B1	60	5.459	14.437
B2	60	6.270	15.434

Table 14.
Wear rate of hybrid composites and AZ91 alloy (dry and wet 60 N).

Sample	Load (N)	Wet/Wear rate (mm3/m)x10–3	Dry/Wear rate (mm3/m)x10–3
AZ91	80	17.895	19.415
A1	80	9.541	14.847
A2	80	11.695	14.438
B1	80	11.698	14.572
B2	80	8.826	14.951

Table 15.
Wear rate of hybrid composites and AZ91 alloy (dry and wet 80 N).

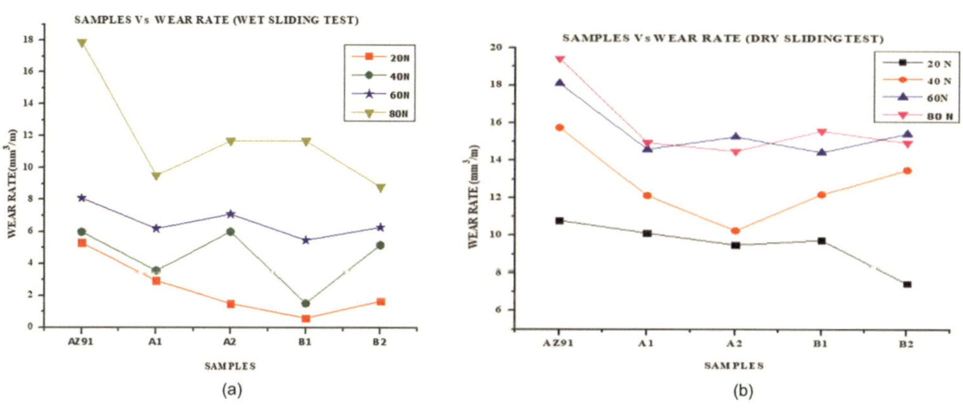

Figure 13.
(a). Samples vs Wear rate for wet sliding Wear test. (b). Samples vs Wear rate for dry sliding wear test.

of the test samples are depicted in **Figure 13** (a) and (b). AZ91 magnesium alloy showed high wear rate both in dry sliding and wet sliding conditions.

9.3 Effect of reinforcement and load

The variation of wear rate as the function of load for all the test samples AZ91, A1, A2, B1, and B2 were plotted as shown in **Figures 14** and **15**. It was noticed that the addition of reinforcement decreased the wear loss of the contacting specimen (pin surface) against hardened steel disc, under all conditions of normal load applied in air.

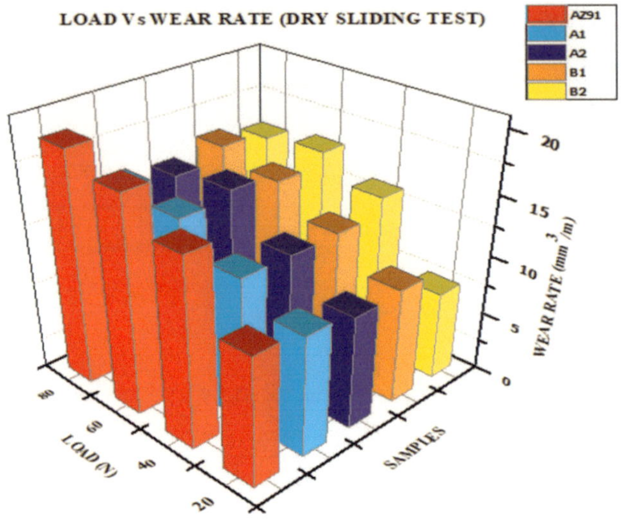

Figure 14.
Load/samples vs Wear rate(x10^{-3}) for dry wear test.

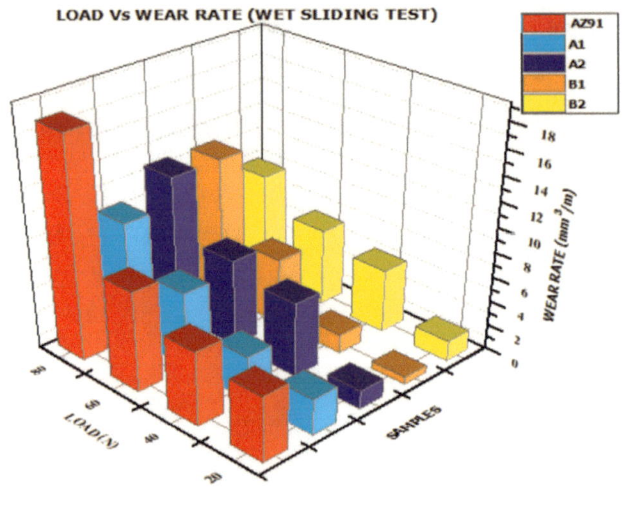

Figure 15.
Load/samples vs Wear rate (x10^{-3}) -wet sliding wear test.

It was found the wear resistance of the hybrid composites showed improved performance than the unreinforced base alloy. This behaviour in the hybrid composites was attributed due to two main reasons are (i) the addition of reinforcements and (ii) refinement in the grain size of the hybrid composites.

The addition of titanium and graphene had added enough hardness to the novel material to resist the wear due to sliding. At higher loads, AZ91/Ti/0.5% Gr showed more evident result in difference in wear rate than AZ91 reference alloy. Within the hybrid composite, varying weight percentage in titanium exhibited better wear resistance. Similar results were reported in the previous work [27]. But the addition graphene along with titanium in the AZ91 base alloy had better results.

The other system AZ91 reinforced with titanium dioxide and graphene revealed less wear loss even at higher load conditions. The presence of titanium dioxide in the matrix along with graphene, acts as solid lubricants protecting the pin sample from the direct contact with the disc surface and also the material loss [28]. Within the hybrid composites, varying weight percentage of titanium dioxide exhibited better wear resistanceAZ91/1% TiO_2/0.5%Gr. At higher loads, it was inferred that the both the combination of reinforcements with base alloy showed similar behaviour of wear loss.

The wear resistance and specific wear resistance each of hybrid composites and AZ91 magnesium alloy are illustrated in the **Tables 16–19** under different load

Sample	Load (N)	Wet/Wear resistance x10^3	Specific wear rate (mm^3/Nm)x10^{-3}	Dry/Wear resistance x10^3	Specific wear rate (mm^3/Nm)x10^{-3}
AZ91	20	0.1889	0.2648	0.0927	0.539
A1	20	0.3423	0.1461	0.0988	0.506
A2	20	0.6702	0.0746	0.1053	0.475
B1	20	0.7094	0.0293	0.1026	0.487
B2	20	0.6098	0.0820	0.1342	0.373

Table 16.
Wear resistance and specific wear rate of hybrid composites and AZ91 alloy (dry and wet 20 N).

Sample	Load (N)	Wet/Wear resistance x10^3	Specific wear rate (mm^3/Nm)x10^{-3}	Dry/Wear resistance x10^3	Specific wear rate (mm^3/Nm)x10^{-3}
AZ91	40	0.1672	0.1495	0.0635	0.3935
A1	40	0.2814	0.0889	0.0825	0.3030
A2	40	0.1670	0.1497	0.0976	0.2563
B1	40	0.6618	0.0378	0.0821	0.3046
B2	40	0.1938	0.1290	0.0740	0.3376

Table 17.
Wear resistance and specific wear rate of hybrid composites and AZ91 alloy (dry and wet 40 N).

Sample	Load (N)	Wet/Wear resistance x10^3	Specific wear rate (mm^3/Nm)x10^{-3}	Dry/Wear resistance x10^3	Specific wear rate (mm^3/Nm)x10^{-3}
AZ91	60	0.1236	0.1348	0.0553	0.3015
A1	60	0.1618	0.1030	0.0685	0.2432
A2	60	0.1411	0.1181	0.0655	0.2543
B1	60	0.1832	0.0910	0.0693	0.2406
B2	60	0.1595	0.1045	0.0648	0.2572

Table 18.
Wear resistance and specific wear rate of hybrid composites and AZ91 alloy (dry and wet 60 N).

Sample	Load (N)	Wet/Wear resistance $x10^3$	Specific wear rate $(mm^3/Nm)x10^{-3}$	Dry/Wear resistance $x10^3$	Specific wear rate (mm^3/Nm) $x10^{-3}$
AZ91	80	0.0559	0.2237	0.0515	0.2427
A1	80	0.1048	0.1193	0.0674	0.1856
A2	80	0.0855	0.1462	0.0693	0.1805
B1	80	0.0855	0.1462	0.0686	0.1822
B2	80	0.1133	0.1103	0.0669	0.1869

Table 19.
Wear resistance and specific wear rate of hybrid composites and AZ91 alloy (dry and wet - 80 N).

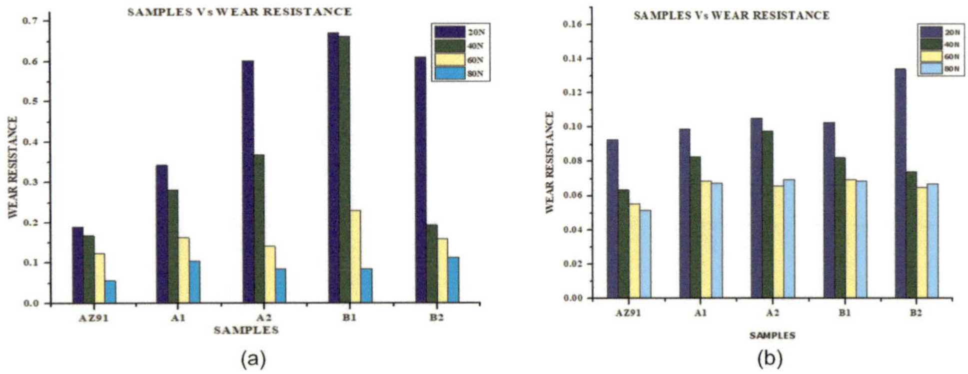

Figure 16.
(a). Samples vs Wear resistance($x10^3$) for wet sliding Wear test. (b). Samples vs Wear resistance($x10^3$) for dry wear test.

condition. It was also inferred that the wear loss was remarkably lower in wet sliding condition than dry sliding as depicted in **Figure 16** (a) and (b). Hence the wear resistance was greatly improved in wet sliding condition.

The reports of earlier researches represent that increase in load, increases the wear loss. In sight about the results clearly depicts that presence of titanium or titanium dioxide with 0.5% graphene as reinforcements for the hybrid composites as hard phase in the magnesium matrix results in strengthening. As a result, it provides improved resistance against plastic deformation.

9.4 Surface analysis of Wear samples

In order to explore the wear behaviour of the novel hybrid composites under different test condition and also the presence of reinforcements nanoparticles effect, wear track of test samples are recorded using optical microscope for dry sliding and wet sliding wear test under the load conditions of 20 N, 40 N, 60 N, and 80 N. As well as SEM analysis was done for clear understanding of wear mechanism. The **Figures 17–24** depicts the wear track of AZ91, hybrid composites A1, A2, B1, and B2in the dry and wet sliding wear condition respectively.

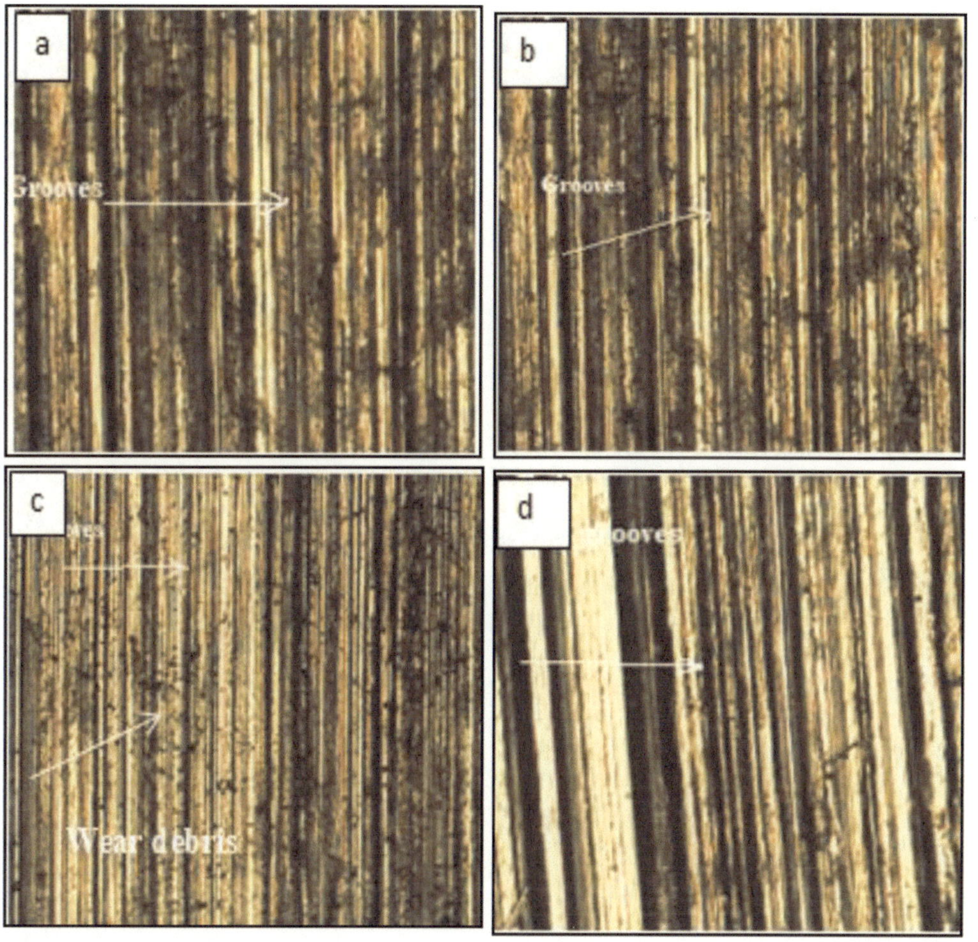

Figure 17.
Optical microscope image of worn surface of AZ91 for load 20 N, 40 N, 60 N, and 80 N (dry condition).

By Archard's law, as the hardness of the material increases, the wear resistance property is greatly improved and wear rate is less. Considering all the test condition, it was found from the worn surface morphology of the test samples that there were three wear mechanism abrasion, delamination and oxidation. Adhesion was not very significantly found in the hybrid composites. This was mainly due to the presence of graphene which acts as solid lubricant. In addition to this, titanium dioxide not only imparts hardness to the materials but also act as lubricant. This may be one of the reasons for wear loss to be least in hybrid composites fabricated in this combination AZ91/TiO_2/0.5%Gr.

The SEM images of the worn surface were shown in **Figures 25-29** in dry and wet condition to evalute the wear behaviour. It was inferred clearly that dry sliding test produced grooves on the pin surface incurring heavy material loss due to ploughing action of harder surface. It was also found that AZ91 magnesium alloy exhibited severe deformation with deep groove and cracks. The titanium and graphene reinforced hybrid composites showed grooves but at high weight percentage the grooves were deeper which depicts either detachment of reinforcement and breaking

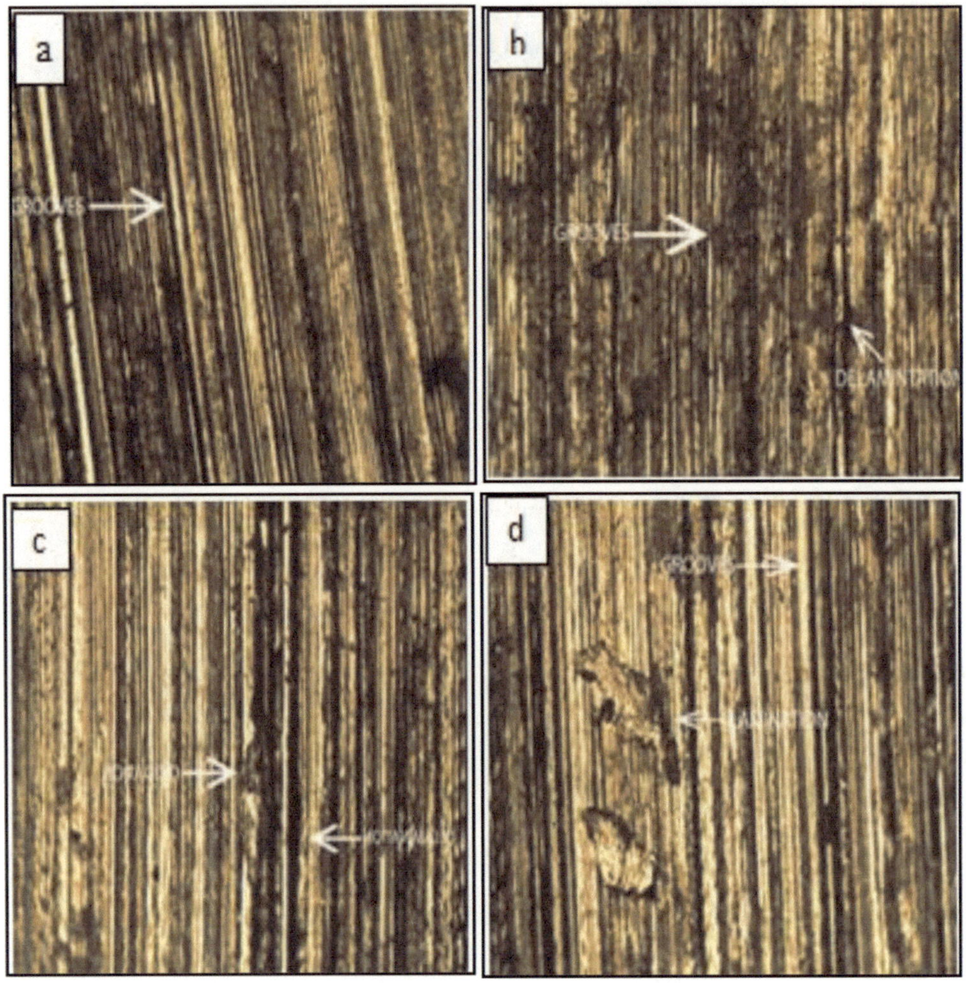

Figure 18.
Optical microscope image of worn surface of AZ91/1%Ti/0.5%Gr for load 20 N, 40 N, 60 N, and 80 N (dry condition).

brittle layer formed. But in titanium dioxide and graphene reinforced hybrid composite showed only fine grooves in dry condition and very fine scratches in wet condition. It evident that hybrid composite reinforced with titanium dioxide and graphene possess better wear resistance.

10. Summary

From the experimental investigation the following results were obtained. The hybrid composites exhibited improved wear resistance due to the addition of the titanium and graphene reinforcements.

Among the combination of hybrid composites, AZ91/1% TiO_2/0.5%Gr and AZ91/2%TiO_2/0.5%Gr exhibited better wear resistance compared to AZ91/x%Ti/0.5%Gr. This due to the presence of stronger TiO_2 dispersed along with graphene act as solid lubricant during the sliding action between the contact surfaces.

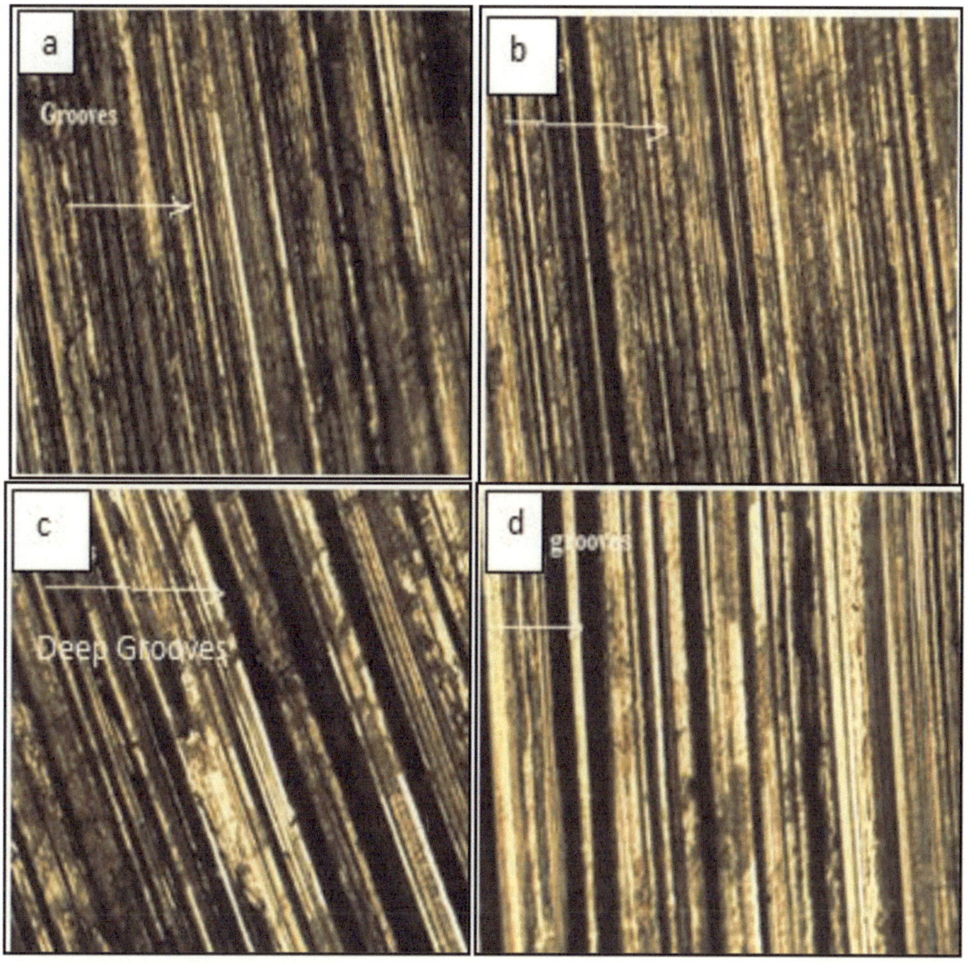

Figure 19.
Optical microscopc image of worn surface of AZ91/2%Ti/0.5%Gr for load 20 N, 40 N, 60 N, and 80 N (dry condition).

11. Conclusion

In the present work, a new and cost effective magnesium based alloy hybrid material was developed successfully by stir casting technique. Addition of reinforcements like titanium, titanium dioxide and graphene has shown significant change in material's property. It was inferred from the results obtained that hybrid composites exhibited improved properties than AZ91 base alloy. The main conclusion obtained from comparison of each combination of reinforcements system are listed below

- AZ91–2%TiO_2–0.5% Gr showed good tensile strength and hardness.
- AZ91–2%Ti-0.5% Gr showed high hardness and denser hybrid material among them.
- AZ91–2%Ti-0.5% Gr exhibited high compressive strength.

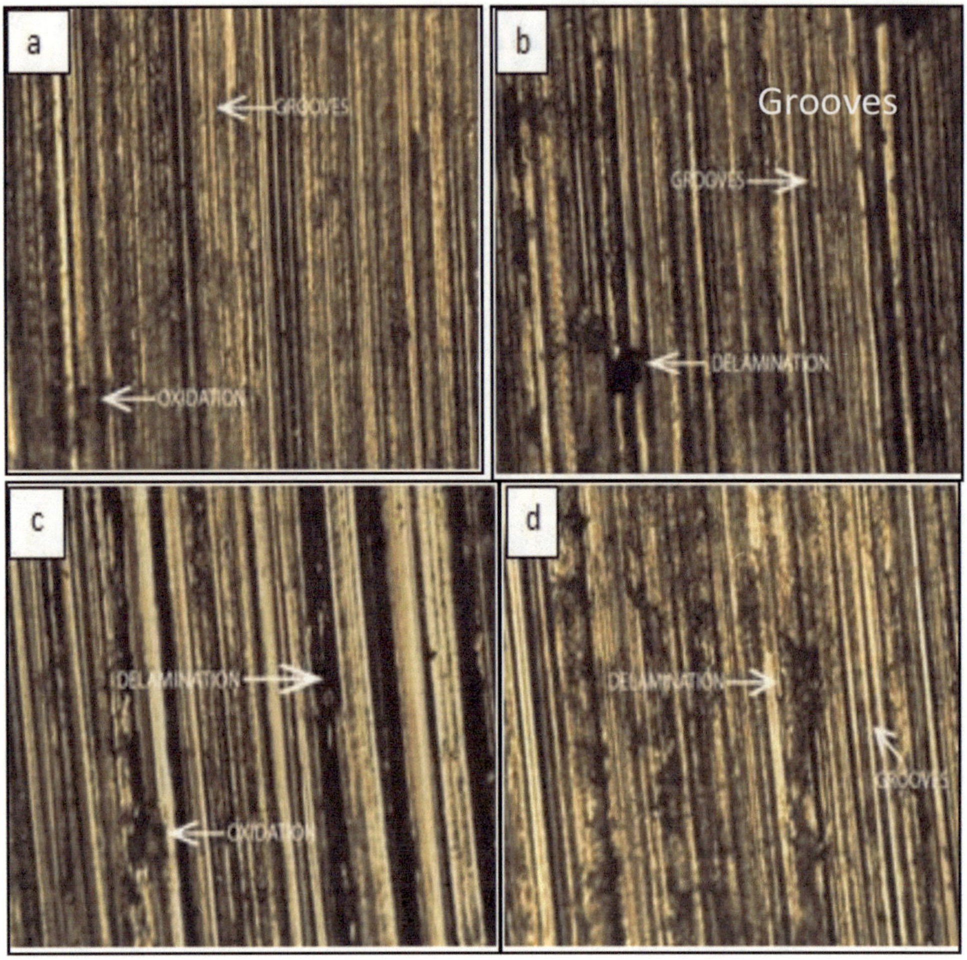

Figure 20.
Optical microscope image of worn surface of AZ91/1%TiO_2/0.5%Gr for load 20 N, 40 N, 60 N and 80 N (dry condition).

Wear resistance of the AZ91 magnesium alloy was greatly enhanced by addition of titanium dioxide and graphene. AZ91–1%TiO_2–0.5% Gr showed better wear property in all the test condition of dry and wet under varying load. Under low load conditions AZ91–2%Ti-0.5% Gr also disclosed better wear resistance and moderate difference at high load. AZ91–2%Ti-0.5% Gr showed high hardness but low wear resistance due to the hard Ti particle detachment may lead three body abrasion phenomenon. However, titanium dioxide has sufficient hardness and lubricant property to protective the material from wear. The order of wear resistance B1 > A2 > A1 > B2 > AZ91.

Corrosion resistance was very significant in titanium and graphene combination. Addition of reinforcement greatly prevented the material degradation by formation of stable protective layer. AZ91–2%Ti-0.5% Gr showed high corrosion resistance in

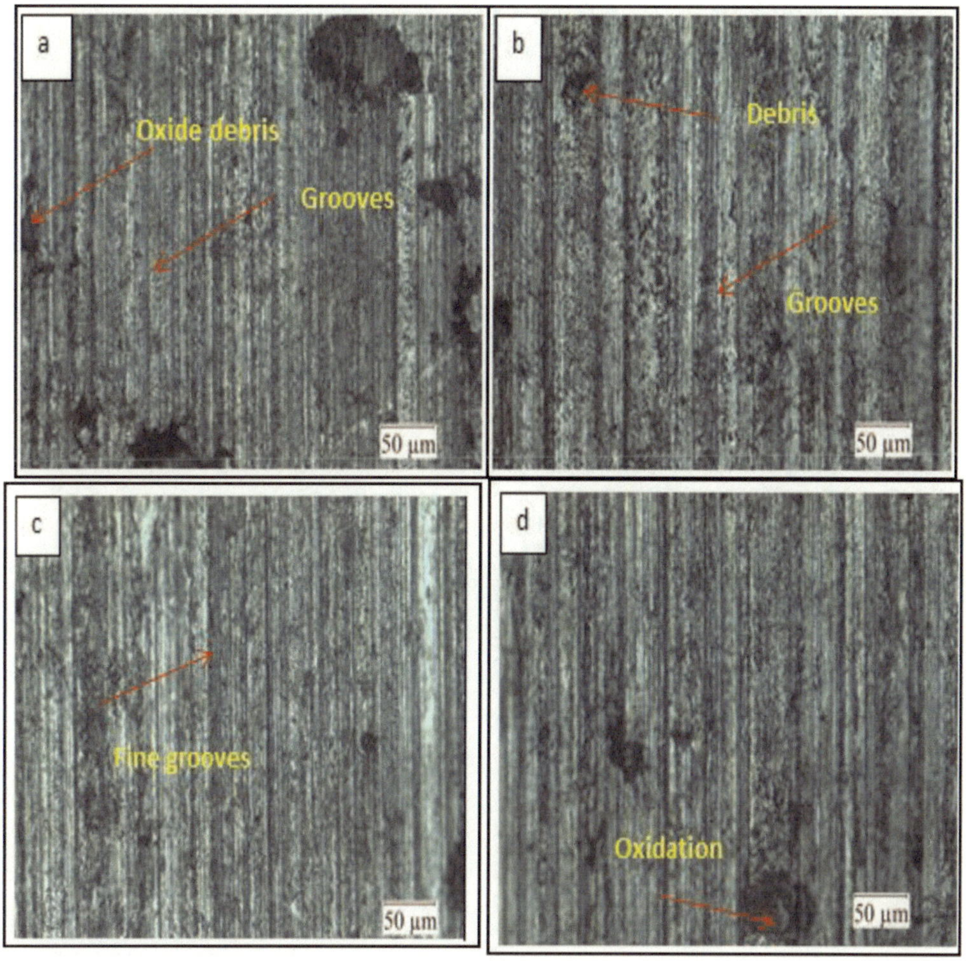

Figure 21.
Optical microscope image of worn surface of AZ91/1%Ti/0.5%Gr for load 20 N, 40 N, 60 N, and 80 N (wet condition).

acidic, neutral and alkaline environment of 3.5% NaCl. The order of corrosion resistance is A2 > A1 > B1 > B2 > AZ91.

Nevertheless all the hybrid composites exhibited better properties than base alloy. Among the four hybrid composites, based on mechanical properties, wear and corrosion behaviour. AZ91–2%Ti-0.5% Gr showed better properties in all aspects.

From engineering application point of view, materials play a key role in design, fabrication, quality, application and cost of the product.

Hence from the present investigation, hybrid composites developed may be considered for the varied field of engineering applications based on the need of usage in different environment like compressive strength or wear resistance or corrosion resistance. Recommended material with improved property based on the property requirement are listed in **Table 20**.

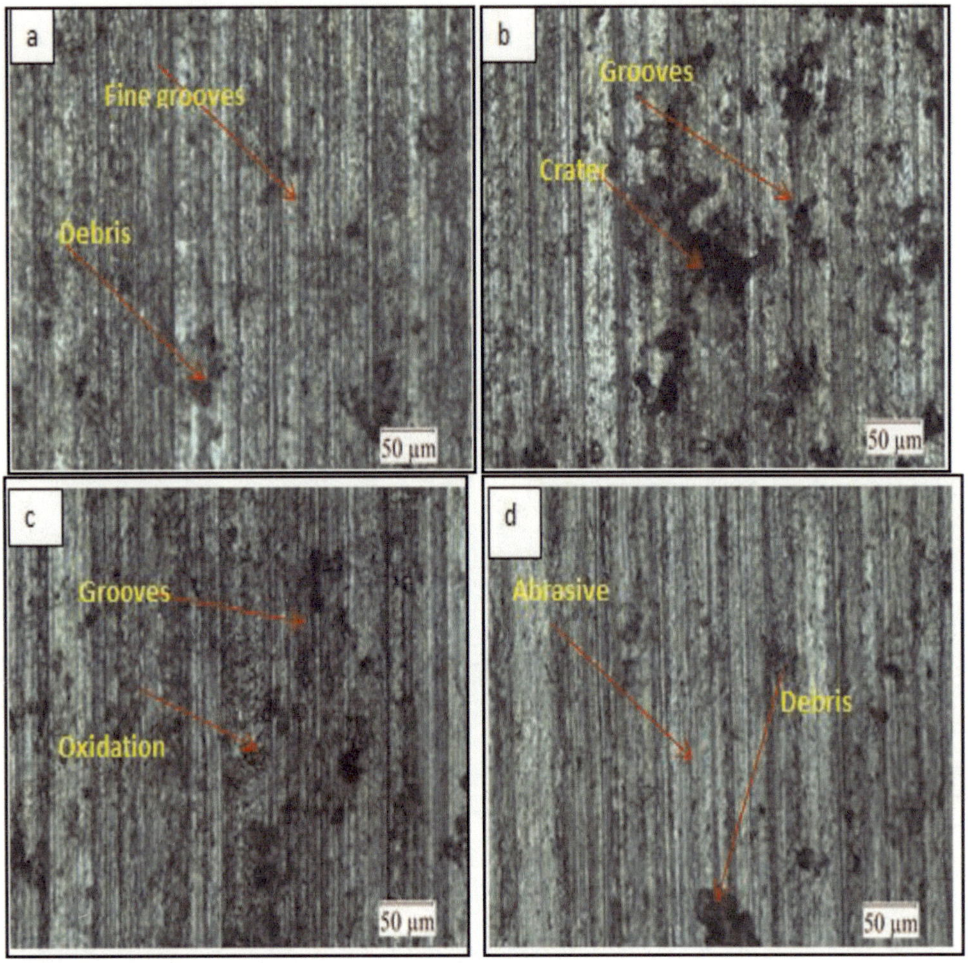

Figure 22.
Optical microscope image of worn surface of AZ91/2%Ti/0.5%Gr for load 20 N, 40 N, 60 N, and 80 N (wet condition).

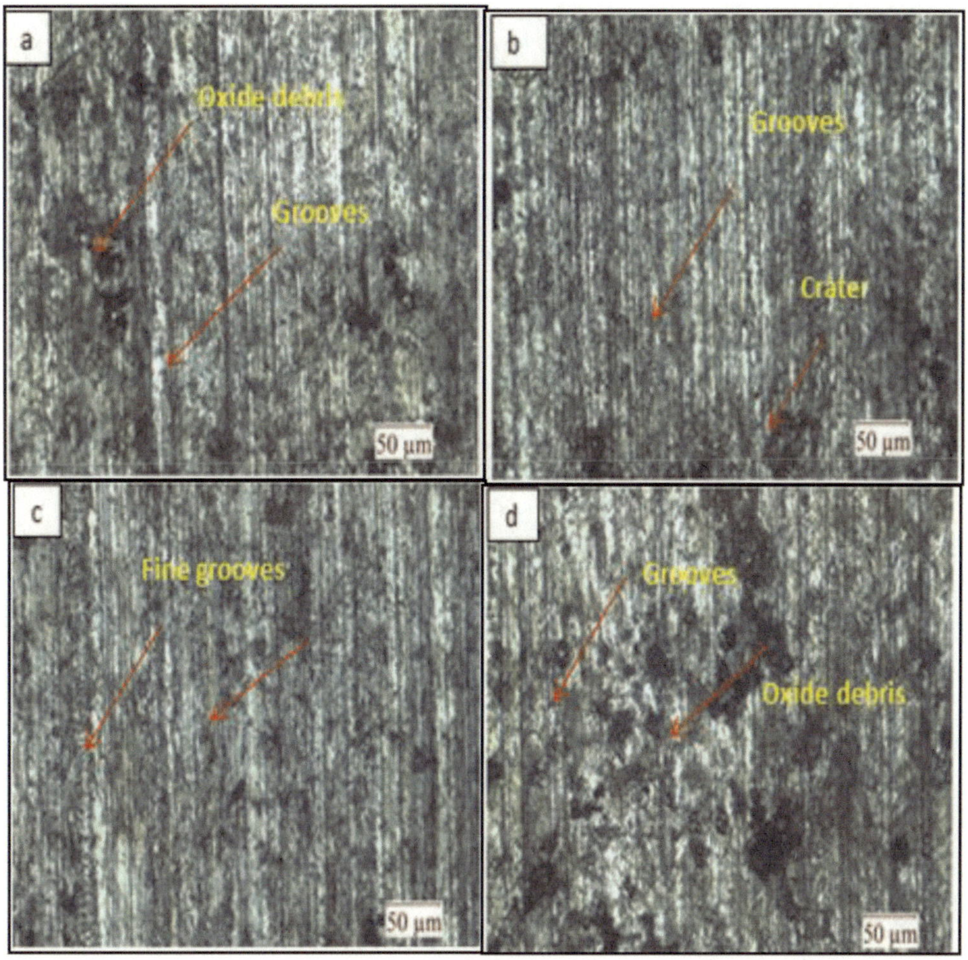

Figure 23.
Optical microscope image of worn surface of AZ91/1%TiO_2/0.5%Cr for load 20 N, 40 N, 60 N, and 80 N (wet condition).

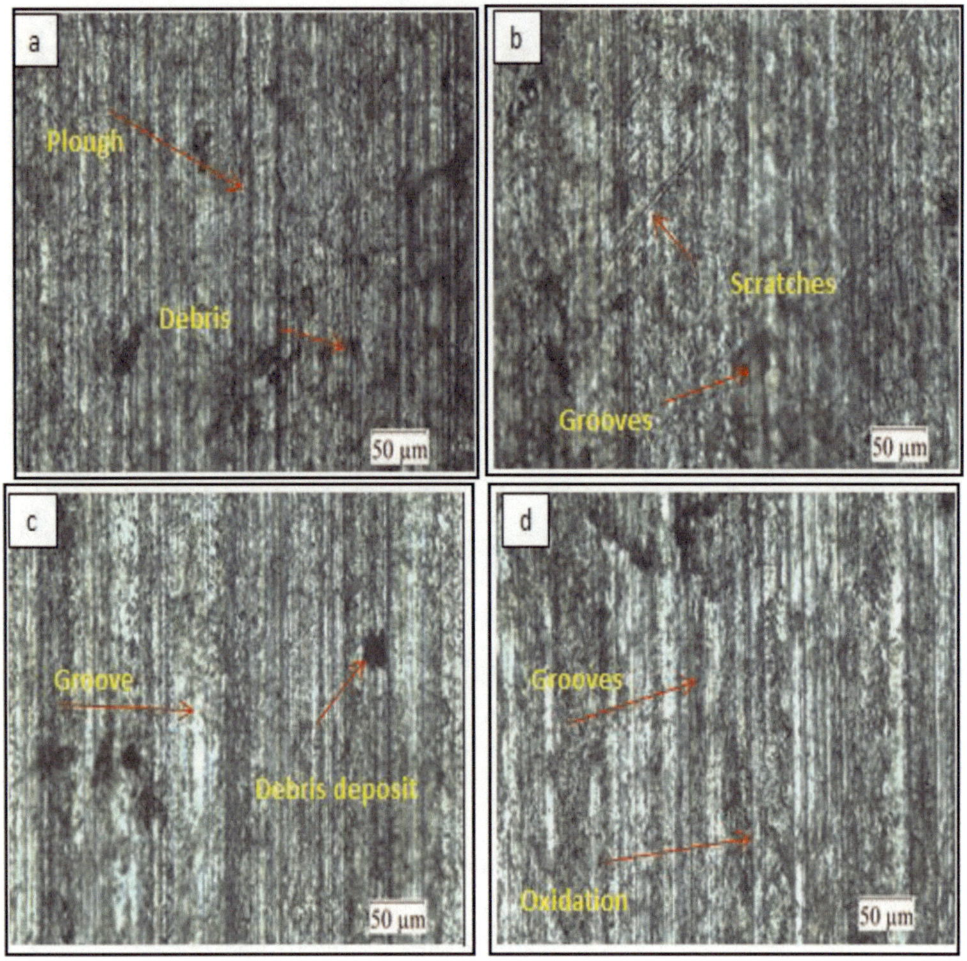

Figure 24.
Optical microscope image of worn surface of AZ91/2%TiO_2/0.5%Gr for load 20 N, 40 N, 60 N, and 80 N (wet condition).

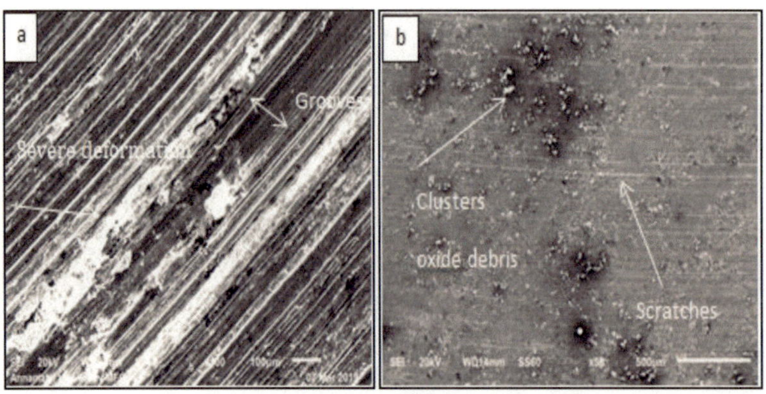

Figure 25.
SEM image of worn surface of AZ91 Mg alloy in (a) dry & (b) wet.

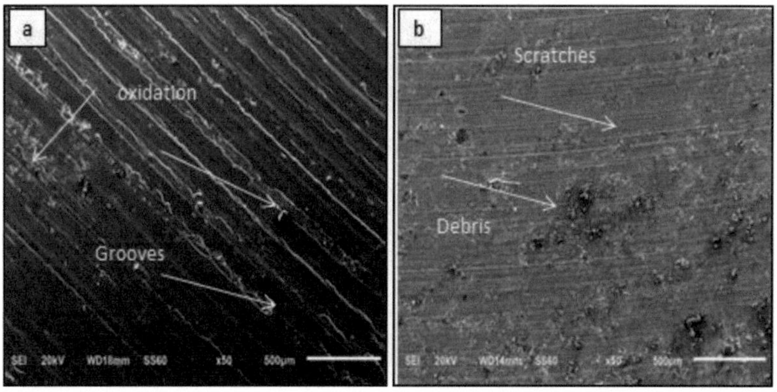

Figure 26.
SEM image of worn surface of AZ91/1%Ti/0.5%Gr (a) dry and (b) wet.

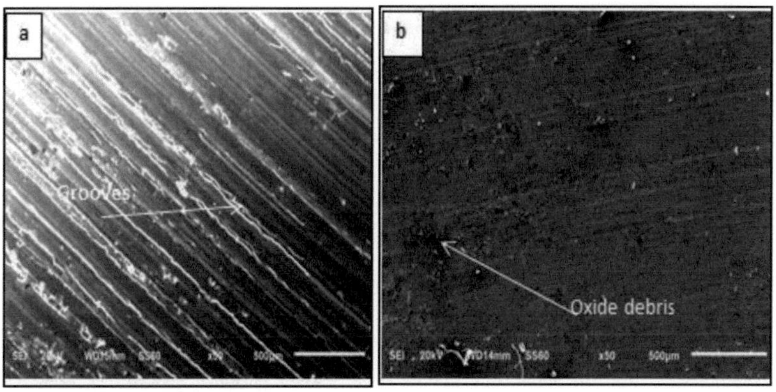

Figure 27.
SEM image of worn surface of AZ91/2%Ti/0.5%Gr (a) dry and (b) wet.

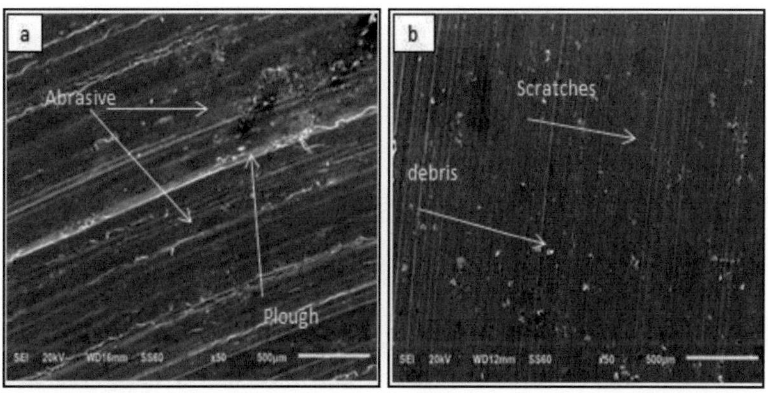

Figure 28.
SEM image of worn surface of AZ91/1%TiO_2/0.5%Gr (a) dry and (b) wet.

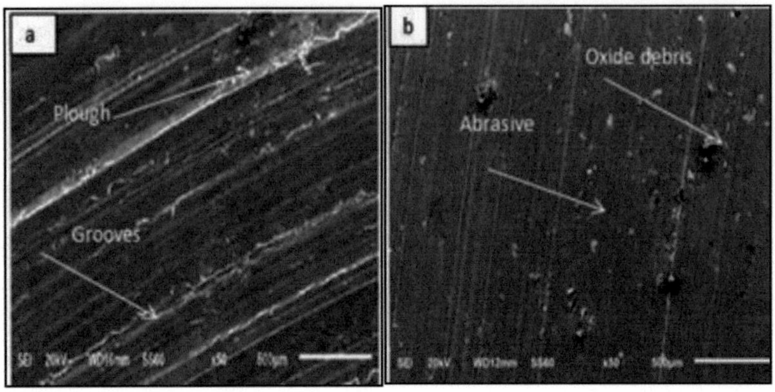

Figure 29.
SEM image of worn surface of AZ91/2%TiO_2/0.5%Gr (a) dry and (b) wet.

Properties requirement	Material Suggested
High compressive strength	AZ91–2%Ti-0.5%Gr AZ91–2%TiO_2–0.5% Gr
High Wear resistance	AZ91–1%TiO_2–0.5%Gr
High Corrosion Resistance	AZ91–2%Ti-0.5%Gr
Moderate strength,Wear and Corrosion	AZ91–1%Ti-0.5%Gr

Table 20.
Material requirement.

Author details

Palanivel Mathiazhagan* and S. Jayabharathy
Puducherry Technological University, Puducherry, India

*Address all correspondence to: pmathi@pec.edu

References

[1] Vinay Vaish and Hitender Mehta. India: Environment Laws In India, The Environment Protection Act, 31 August, 1986. 2017

[2] Mordike BL, Ebert T. Magnesium, properties-applications-potential. Materials Science and Engineering: A. 2001;**302**(1):37-45

[3] Blawert C, Hort N and Kainer KU. Automotive applications of magnesium and its alloys. Trans. Indian Inst. Met. 2004;**57**(4):397-408

[4] Sameer Kumar D, Suman KNS. Selection of Magnesium alloy by MADM Methods for Automobile Wheels. International Journal of Engineering and Manufacturing. 2014;**4**(2):31-41

[5] Tiancai X, Yang Y, et al. Overview of advancement and development trend on magnesium alloy. Journal of Magnesium and alloys. 2019;**7**:536-544

[6] Kainer KU, Von Buch F, The Current state of technology and Potential for further development of Magnesium application, Magnesium Alloys and Technology, Edited by K.U. Kainer. Wiley online library; Feb 2003. pp. 1-22. ISBN: 3-527-305704

[7] Fink R. Oskar Frech Gmb H, Schorndorf, Die Casting Magnesium, Magnesium –alloy and technology, edited by K.U. Kainer. Wiley online library; 2003. pp. 23-24. ISBN: 3-527-305704

[8] Sankar N, Chandramohan V. Dr. M. Eswaramoorthi, investigation of lightweight materials used in automobile applications. International journal of Engineering Development and research. 2016;**4**(1):2321-9939

[9] Andrej Athens, Guang–Ling Song, Fuyong Cao, Zhiming Shi, Patrick K. Bowen. Advances in Mg corrosion and research suggestion. Journal of Magnesium and Alloys. 2013;**1**(3): 117-200

[10] Dey A. Krishna Murari Pandey, magnesium metal matrix composite- a review. Reviews on Advanced MaterialsScience. 2015;**42**:58-67

[11] Alan A. Luo. Magnesium casting technology for structural applications. Journal of Magnesium and Alloys. 2013; **1**(1):2-22

[12] Aghion E, Bronfin B, Eliezer D. The role of magnesium industry in protecting the environment. Journal of Materials Processing Technology. 2001; **117**:381-385.

[13] Tharumarajah A, Koltun P. Is there an environmental advantage of using magnesium components for light weighting cars?. Journal of Cleaner Production. 2007;**15**(11-12):1007-1013

[14] Musfirah A. H, Jaharah A. G. Magnesium and Aluminium alloy in Automotive Industry. Journal of Applied Sciences Research. 2012;**8**(9):4865-4875

[15] Zuzanka, Trojanova, Zoltan Szaraz et al, Magnesium Alloy Based Composites, Magnesium alloys-Design, Processing and Properties. In: Frank Czewinski, editors. Intech open access publisher; Jan 2011. ISBN: 987-953- 307-520-4

[16] Aravindan S, Rao PV, Ponappa K. Evaluation of physical and mechanical properties of AZ91D/SiC composites by two step stir casting process. Journal of Magnesium and alloys. 2015;**3**:52-62

[17] Ch.Fritze, BMW Group, Munchen. Fibre-reinforced magnesium composites, magnesium alloys and technology, edited by K.U. Kainer. 2003 ISBN: 3-527-305704

[18] Yu Z, Tang A, Zhang L, Pan F. Effect of micro alloy with titanium on microstructure and mechanical properties of AZ91 magnesium alloy. Material Science and Tehnology. 2014; **30**(12):1441

[19] Lakshmanan Pillai A, Jinu GR. Synthesis of titanium oxide particles reinforced with magnesium by argon controlled stir casting process and characterization. Chemical and Materials Engineering. 2017;**5**(2):26-33

[20] Soorya Prakash K, Balasundar P, Nagaraja S, Gopal PM, Kavimani V. Mechanical and wear behaviour of Mg-SiC–gr hybrid composites. Journal of Magnesium and Alloys. 2016;**4**(3): 197-206

[21] Handbook ASM. Volume 8 Mechanical Testing and Evaluation. ASM Headquarters and Geodesic Dome, Materials Park campus, Russell Township, Geauga County, Ohio: ASM international; 2003

[22] Rashad M, Pan F, Asif M. Xianhua chen, improved mechanical properties of magnesium based composite with titanium-aluminum hybrids. Journal of Magnesium and Alloys. 2015;**3**(1):1-9

[23] Candan S, Unal M, Koc E, et al. Effects of titanium addition on mechanical and corrosion behaviours of AZ91 magnesium alloy. Journal of Alloys and Compounds. 2011;**509**(5):1958-1963

[24] Kandemir S. Development of graphene Nanoplatelets reinforced AZ91 magnesium alloy by solidification processing. Journal of Materials Engineering and Performance. 2018; **27**(6):3014-3023

[25] Aatthisugan I, Razal Rose A, Selwyn Jebadurai D. Mechanical and wear behaviour of AZ91 magnesium matrix hybrid composite reinforced with boron carbide and graphite. Journal of Magnesium and Alloy. 2015;**5**(1):20-25

[26] Nithin NS, Venkit H. Investigation of Wear analysis of AZ91 magnesium alloy reinforced with TiC & BN. IJETT. 2015;**28**:236-242

[27] Vikram KS, Jain JV, Muthukumaran S. Effect of first and second passes of titanium dioxide – reinforced Aluminium surface composite via friction stir processing. Arabian Journal for Science and Engineering, Springer. 2018

[28] Jayabharathy PM. Investigation of Mechanical & Wear behaviour of AZ91 magnesium matrix hybrid composite with TiO2/graphene. Materials Today, Elvesier. October 2019:2394-2397

Chapter 6

Magnesium Borates: The Relationship between the Characteristics, Properties, and Novel Technologies

Fatma Tugce Senberber Dumanli

Abstract

Magnesium borates are compounds including mainly magnesium (Mg), boron (B) oxygen (O), and hydrogen (H). Magnesium borates are traditionally famous for their strong thermoluminescence, mechanical and thermal features due to their high elasticity coefficient, corrosion, and heat resistance. Because of being beneficial, especially in the applications such as thermoluminescence and X-ray screening, and ease of synthesis, magnesium borates are produced by using different experimental procedures exhibiting different characteristics. Main traditional synthesis techniques can be classified as liquid state and solid-state synthesis methods. With the help of novelties in synthesis technology, new techniques are beginning to emerge in magnesium borate syntheses such as hybrid synthesis, ultrasound, microwave, and capping agent addition. The strengthened characteristics of the compounds would lead to new applications such as stomach cancer chemotherapy and wastewater treatment. In this chapter, it is aimed to make a comparison between the characteristics of synthesized magnesium borates and their properties. In addition, new types of magnesium borates obtained by various synthetic techniques are expected to be useful for industrial applications such as space technology, radiation dosimetry, X-ray screening, ion batteries, and hydrocarbon reaction catalysis. Such classification of properties and the synthesis techniques will enlighten the relationship between the characteristics and novel applications of magnesium borates.

Keywords: advanced synthesis methods, characterization, magnesium borates, microwave synthesis, reaction mechanism, ultrasonic synthesis

1. Introduction

Magnesium is the third most abundant element by mass in Earth's surface composition. Magnesium-based materials are preferable, especially for their lightweight. Magnesium is found in nature as the combination of oxygen to form different minerals such as sulfates, carbonates, nitrates, and borates [1, 2]. Some kinds of magnesium compounds can be soluble, such as magnesium sulfate, magnesium nitrate, and magnesium bromide. These compounds are generally hygroscopic. Other types of magnesium compounds are known as insoluble, such as magnesium borate, magnesium oxide, and magnesium phosphate. The salt group of magnesium is effective in the corrosive behavior of magnesium compounds [3]. Being insoluble in water makes magnesium borates production easier with developed technology, such as hybrid synthesis methods, the use of microwave and ultrasound technologies.

As a magnesium-based compound, magnesium borate can be utilized in the applications of X-ray screening, radiation permeation, catalysis of organic reactions, strengthening of plastics, and ion-battery systems. The studies on magnesium borates are generally focused on synthesis techniques. The common aim of the developed synthesis methods is to decrease energy consumption. However, the characteristics of the prepared sample are related to the designed experimental setup. A thorough understanding of the relationship between synthesis procedures and characteristics of the novel borates obtained will help increased the use of the correct form of magnesium borate in industry.

2. Magnesium borate minerals

As being notable magnesium compounds, magnesium borates mainly include the atoms of Mg, B, O; however, other types of elements may be included according to the reserves they are mined. According to the conditions they formed in nature or fabricated in the laboratory, the magnesium borates can include crystal waters and/or hydroxyl groups. Therefore, this type of magnesium minerals can be classified as hydrated or dehydrated forms. The common examples of identified magnesium borates and their crystal systems are presented in **Table 1**.

Type	Mineral name	Chemical formula	Crystal system
Hydrated	Admontite	$MgB_6O_{10}{\cdot}7H_2O$	Monoclinic
	Aksaite	$Mg[B_6O_{10}(OH)_6]{\cdot}2H_2O$	Orthorhombic
	Halurgite	$Mg_4[B_8O_{13}(OH)_2]_2{\cdot}7H_2O$	Monoclinic
	Hungchaoite	$MgB_4O_7{\cdot}9H_2O$	Triclinic
	Hydroxylborite	$Mg_3(BO_3)(OH)_3$	Hexagonal
	Inderite	$MgB_3O_3(OH)_5{\cdot}5H_2O$	Monoclinic
	Kurnakovite	$MgB_3O_3(OH)_5{\cdot}5H_2O$	Triclinic
	Mcallisterite	$Mg_2[B_6O_7(OH)_6]_2{\cdot}9H_2O$	Trigonal
	Pertsevite-(OH)	$Mg_2(BO_3)(OH)$	Orthorhombic
	Pinnoite	$Mg[B_2O(OH)_6]$	Tetragonal
	Preobrazhenskite	$Mg_3B_{11}O_{15}(OH)_9$	Orthorhombic
	Szaibélyite	$MgBO_2(OH)$	Monoclinic
	Wightmanite	$Mg_5(BO_3)O(OH)_5{\cdot}2H_2O$	Monoclinic
Dehydrated	Kotoite	$Mg_3[BO_3]_2$	Orthorhombic
	Suanite	$Mg_2[B_2O_5]$	Monoclinic

Table 1.
Identified magnesium borates and their crystal systems [4].

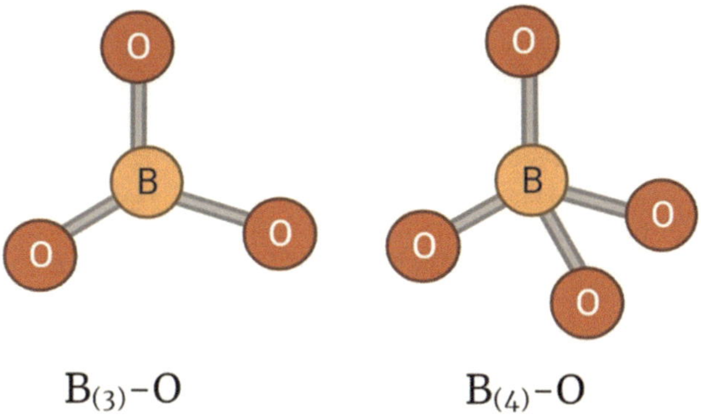

Figure 1.
The examples of boron-oxygen linkages $B_{(3)}$-O and $B_{(4)}$-O.

The arrangement of functional groups in the molecule determines the properties of magnesium borate [5]. As a chemical compound, magnesium borates mainly include the functional groups of three and four coordinated borate anions ($B_{(3)}$-O and $B_{(4)}$-O) connected to the magnesium atoms (**Figure 1**). The typical symmetric and asymmetric stretching in a molecule can be determined by Fourier-transform infrared spectroscopy (FT-IR) and/or Raman Spectroscopy.

Spectral analyses result of Admontite ($MgB_6O_{10}{\cdot}7H_2O$) samples prepared at different reaction times in hydrothermal conditions were presented in **Figure 2**. In FT-IR analyses of magnesium borates, the spectrum begins with the peak around 3500 cm^{-1} which indicates the crystal water for the hydrated compounds. The region above the 1600 cm^{-1} is generally called as "free H_2O zone". The effects of hydroxyl anions are seen as "bending of hydroxyl groups in plane" and "bending of hydroxyl groups out of plane". The peaks between 1400 and 1200 cm^{-1} indicate "bending of hydroxyl groups in plane" whereas the peaks between 950 and 750 cm^{-1} are related with the "bending of hydroxyl groups out of plane". The stretching between boron and oxygen atom is commonly seen between 1600 and 650 cm^{-1}. The peaks in the region of 1600–1400 cm^{-1} are related with the "asymmetric stretching of $B_{(3)}$-O". "Asymmetric stretching of $B_{(4)}$-O" can be explained with the peaks in the range of 1200–950 cm^{-1}. The peaks at the lower wavelength values of 750 cm^{-1} can be interpreted with the "bending of $B_{(3)}$-O" [6–8].

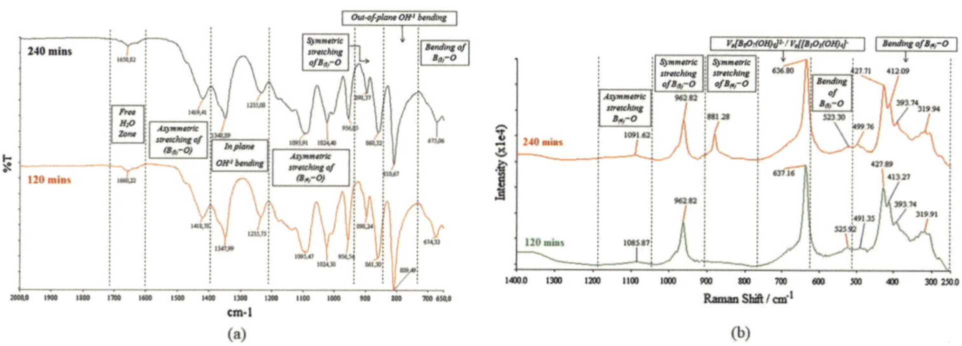

Figure 2.
Spectral analyses result of Admontite prepared at different reaction times (a) FT-IR spectra, and (b) Raman spectra [6].

In Raman analyses of magnesium borates, the characteristic peaks are seen in the wavelength region of 1200–250 cm^{-1}. The peaks between 1200 and 1050 cm^{-1} are interpreted with the "asymmetric stretching of $B_{(4)}$-O". "Symmetric stretching of $B_{(3)}$-O" and "symmetric stretching of $B_{(4)}$-O" are seen in the range of 1050–900 cm^{-1} and 900–750 cm^{-1}, respectively. For the hydrated forms of magnesium borates, the stretching for polyanion of $[B_6O_7(OH)_6]^{-2}$ and $[B_3O_3(OH)_4]^{-2}$ is seen between 750 and 620 cm^{-1}. The "bending of $B_{(3)}$-O" and "bending of $B_{(4)}$-O" can be seen in the Raman shift wavelength ranges of 620–500 cm^{-1} and 500–250 cm^{-1}, respectively [6, 7, 9].

2.1 Properties

The characterization studies on magnesium borates make them preferable in industrial applications. These compounds are known for their superior thermal and mechanical strength, stability, and high coefficient elasticity. According to their specific characteristics, magnesium borates can be used as anticorrosive agent, catalyst, lubricant, and adsorbent. Due to their thermoluminescence properties, they also have applications in radiation dosimetry, X-ray screens, space research, and nonlinear optic laser systems [10–16].

Magnesium borates can be utilized in hydrogen storage systems, acoustic insulation, and ion-battery systems thanks to their high corrosion resistance. The viscosity of melted magnesium borates is relatively low and exhibits excellent electroconductivity. Therefore, magnesium borates can be used as either a coating agent on lithium-ion batteries or as an additive for electrolyte solutions [17–20].

For being of biocompatible properties of magnesium, magnesium borate compounds have also begun to be evaluated as a biomaterial for being unhazardous to the environment and human health. The eco-friendly behavior of these compounds increased their applications in health and wastewater treatments. Fan et al., studied the role of magnesium borate on stomach cancer chemotherapy as a hydrogen release agent [21]. Ma and Liu [22] experimented with the Congo Red adsorption from wastewater by using the sample of $2MgO{\cdot}B_2O_3{\cdot}H_2O$ [22].

2.1.1 Thermoluminescence

In nuclear research, dehydrated forms of magnesium borates such as MgB_4O_7, $Mg_2B_2O_5$, and MgB_2O_4 are generally preferred. This can be explained by the decreasing hygroscopicity at reaction temperatures higher than 950°C and the ease of solid-state synthesis methods [11, 23–25]. Souza et al. [26] compared the thermoluminescence features of the synthesized magnesium borates in liquid-state and solid-state conditions and indicated better results of dehydrated samples [26]. MgB_4O_7 is the most studied composition among the magnesium borates, due to its thermoluminescence behavior. The studies on the thermoluminescence behavior of magnesium borates indicated their suitability of them in beta, neutron, and radiation dosimetry. Several rare earth elements of Cerium, Dysprosium, Samarium, Silver, Terbium, Thulium, have been doped to increase their efficiency in applications [23, 26–29]. Also, Prokic and Christeen [30] and Pellicioni et al. [31] experimented with the beneficial effects of graphite addition to magnesium borates; and the 3% graphite content was determined suitable for the optimum thermoluminescence [30, 31].

2.1.2 Mechanical strength

Modification of thermoplastic materials with dehydrated magnesium borates can strengthen the tensile strength and strain failure. This situation can be explained with the increased physical crosslinking density and decrease in the size of bubble growth. For the mechanical strength increase, $Mg_2B_2O_5$ is commonly preferred in literature [32]. Zhang et al. [33], analyzed the strengthening effects of magnesium borate addition on aluminum-based composites [33]. Baghebanadi et al., indicated the beneficial effects of dehydrated magnesium borates ($Mg_2B_2O_5$ and $Mg_3B_2O_5$) in addition to the cold crushing strength of magnesium-graphite composite [34].

2.1.3 Catalyst effect in reactions

Catalyst effect of magnesium borates can be utilized to both increase reaction conversion in the reactions of hydrocarbon and/or they can also be evaluated to catalyze the other types of inorganic borates such as boron nitride [35, 36]. In the catalytic utilization of magnesium borate, the purity and the morphology of prepared magnesium borate are notable. In this case, the synthesis of magnesium borate at different morphologies will promote the comprehensive use of this type of compound.

Ahmad et al. [10], studied the catalyst effect of rod-like magnesium borates on the electrochemical activity of the dopamine enzyme [10]. Intemann et al. [13], used magnesium borates to catalyze the selective reduction of pyridine [13]. Loiland et al. [35], investigated the catalysis effect of magnesium borate complexes on the oxidative dehydrogenation of ethane and propane mixtures [35].

2.1.4 Adsorption behavior

The determination of the adsorption behavior of magnesium borates is an up-and-coming practice among its applications. The few researches on the adsorption behavior of these compounds include the azo anionic dye of Congo red adsorption on the hierarchic porous particles of magnesium borates. According to the isothermal and kinetic estimations of the adsorption study, the adsorption mechanism can be explained with the Langmuir isotherm and Pseudo second-order kinetic model. The results also showed that adsorbents can be recycled with calcination at 400°C [22, 37, 38]. The comparison of maximum adsorbent capacity values (q_M) for magnesium borates is presented in **Table 2**.

In the studies of Zhang et al. [22] and Ma and Liu [37] magnesium borates were fabricated in hydrothermal conditions whereas Guo et al. [38] preferred thermal conditions [22, 37, 38]. As it is seen in **Table 2**, it has been observed that hydrated compounds have a larger BET surface area and maximum adsorbent capacity. The results indicated that both hydrated and dehydrated forms of magnesium borates can be a promising candidate for toxic dye adsorption.

2.1.5 Thermal behavior

Determination of thermal behavior for the magnesium borates could increase the evaluation probability as fire-retardant agents. Zhang et al. [39] studied the fire retardant effects of magnesium borate addition to the polyvinyl chloride (PVC) and lignin composite [39].

Adsorbent	Morphology	S_{BET} (m^2/g)	q_M (mg/g)	Reference
$2MgO{\cdot}B_2O_3{\cdot}H_2O$	Hierarchic porous	93.46	183.15	[22]
$Mg_2B_2O_5$	Hierarchic porous	24.20	139.30	[37]
$MgBO_2(OH)$	Hierarchic porous	57.22	228.30	[37]
$7MgO{\cdot}2B_2O_3{\cdot}7H_2O$	Hierarchic porous	103.62	202.84	[38]
β-$3MgO{\cdot}B_2O_3$	Hierarchic porous	46.10	170.07	[38]

Table 2.
Comparison of maximum adsorbent capacity values for magnesium borates.

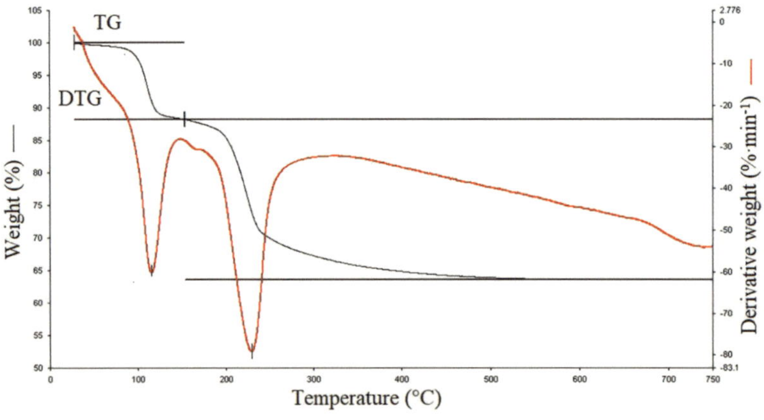

Figure 3.
TG and DTG curves of Admontite mineral ($MgB_6O_{10}{\cdot}7H_2O$).

Thermal behavior of magnesium borates is related to hydrate groups in the structure. For the hydrated magnesium borates, thermal decomposition process begins with the dehydration reaction which means to split off the crystal water ($\cdot xH_2O$), and continues with the dehydroxylation reaction which means to split off hydroxyl anions (OH^-) as a water molecule. With the increasing temperature, phase changes can also be seen.

As a typical example, thermal curves of TG and DTG for Admontite mineral between 25 and 750°C are presented in **Figure 3**. Admontite ($MgB_6O_{10}{\cdot}7H_2O$) mineral has lost its 7 moles of crystal water with a two-step reaction. In the first step, the reaction occurs in the range of 40–125°C, and the peak of DTG curve is seen at 116°C. The second step of decomposition emerges in the range of 125–570°C and the DTG peak is seen at 230°C. The mass losses of the first and second steps of decomposition are determined as 11% and 25%, respectively.

For the dehydrated magnesium borate, only phase changes can be seen or the compound stays stable. This stability could be related with the reaction temperature of fabricated dehydrated magnesium borate.

3. Synthesis procedures of magnesium borates

The characteristic features of the samples associated with the composition, crystalline phases, and morphology are dependent on synthesis procedures. The magnesium borates can be fabricated in various morphologies of the rod, sphere, tube, whisker, belt, wire, porous, or multi-angular at both nanoscale and microscale [15, 40]. Examples of the different morphologies of magnesium borates were presented in **Figure 4**. Kumari et al. [15], synthesized the nano-scale whiskers of magnesium borates in hydrothermal conditions without a capping agent [15]. Liu et al. [40], prepared sub-micron rods of a dehydrated form of magnesium borates by calcination at higher temperatures than 600°C [40]. Guo et al. [38], designed a hybrid method to fabricate the 3D hierarchical flower-like particles of magnesium borates [38].

The experimental design should be both low-cost and eliminate the risk of byproduct formation. The design can be shaped according to the required features of particles. Therefore, synthesis procedures can be classified as liquid-state, solid-state, and hybrid synthesis with the effect of development in production technologies.

3.1 Liquid-state (hydrothermal) synthesis

Liquid-state synthesis of magnesium borates principally includes the dissolution of raw materials in a suitable solvent medium and the reaction occurs with the impulsive effect of temperature increase. Commonly, the type of magnesium salts such as magnesium oxide (MgO), magnesium chloride ($MgCl_2$), magnesium sulfate ($MgSO_4$), and magnesium nitrate ($Mg(NO_3)_2$) is reacted with boric acid (H_3BO_3) or tincal ($Na_2B_4O_7{\cdot}10H_2O$). At the end of the reaction, the solution is filtrated and dried. The growth mechanism could be explained by dissolution, nucleation, and recrystallization.

The growth mechanism according to the study of Ma and Liu is explained in Eqs. (1) and (2) [22]:

$$B_5O_6(OH)_4^{\ -} + H_2O \rightarrow 2B_2O_4(OH)_2^{\ 4-} + H_3BO_3(aq) + 7H^+ \tag{1}$$

$$2Mg^{2+} + B_2O_4(OH)_2^{\ 4-} \rightarrow Mg_2\left[B_2O_4(OH)_2\right] \tag{2}$$

Figure 4.
Examples of the different morphologies of magnesium borates (a) nano-scale whisker by Kumari et al. [15], (b) sub-micron rod by Liu et al. [40], and (c) flower-like particle by Guo et al. [38].

The particle shape and sizes can be controlled by optimizing the liquid-state reaction conditions. Derun et al. [6], fabricated the multi-angular particles of magnesium borate hydrates between the reaction temperatures of 80 and 100°C by using a traditional liquid-state method.

With the developing technology, liquid-state synthesis techniques can also be modified by the use of sonochemistry and capping agents.

3.1.1 Advantages of sonochemistry in liquid-state conditions

As being new and fast-growing technology, ultrasonic treatment has become a significant step in several industrial applications. It is seen that acoustic waves usage shortens the reaction time and increases the reaction yield in comparison with the traditional methods. Due to these effects of cavitation, energy saving can be obtained according to the experimental setup. For this reason, the use of an ultrasonic beam is preferred to obtain proper particle formation.

In many synthesis procedures, the ultrasonic treatment is accepted as a vital step. The employment of ultrasound in the synthesis procedure accelerates the reaction rate and yield. The effects of ultrasound can be operated by factors of power level, cycle, ultrasonic treatment time, and type of ultrasonic reactor. The common types of power sources for ultrasonic treatment in laboratory scale are presented in **Figure 5**.

The reaction mechanism of sonochemistry has not been defined in a detailed way. However, the beneficial contributions of ultrasound make it frequently employable in applications. The studies on the use of sonochemistry are generally based on the prevention of by-product formation, efficient use of raw materials, eco-friendly solvent usage, better waste management (selectivity), and energy savings [41].

Yildirim et al. [42] synthesized the mixtures of Admontite ($MgB_6O_{10}{\cdot}7H_2O$) and Mcallisterite ($Mg_2[B_6O_7(OH)_6]_2{\cdot}9H_2O$) at higher reaction yields than 84% by using

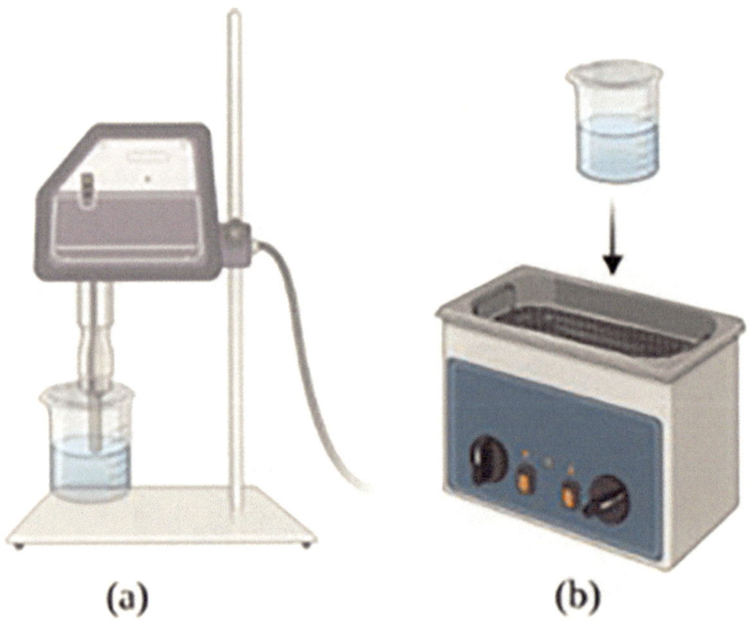

Figure 5.
The examples of ultrasound sources in laboratory scale; (a) ultrasonic prob, and (b) ultrasonic bath.

acoustic cavitation [42]. Kipcak et al. [43], produced the magnesium borate hydrates at higher crystallinity in a sub-micron scale with the effect of ultrasound energy [43]. The comparison of the Admontite morphologies synthesized by the ultrasonic liquid-state method and the traditional liquid-state method was presented in **Figure 6**. As is seen in **Figure 5**. In comparison with the traditional liquid-state methods, smaller particle sizes were observed in the use of ultrasonic-assisted synthesis techniques [6, 42].

3.1.2 Effect of capping agent in liquid-state conditions

The capping agents can be employed to overcome the drawbacks of the synthesis procedures and to produce particles with homogenous and novel morphologies. The selected surfactant can be cationic, anionic, or non-ionic. In the use of capping agent, the type of capping agent is not able to highlight the relationship between the agent and the core particle. Some type of capping agents use could decrease the particle size; however, the crystallinity of samples could be affected adversely [15]. This situation might require a more detailed examination of the relationship between the capping agent and the magnesium borate particle.

In the modified liquid-state synthesis of magnesium borates, the examples of the preferred capping agent are polyvinyl pyrrolidone (PVP), sodium dodecyl sulfate (SDS), nickel nitrate ($Ni(NO_3)_2$), cetyltrimethylammonium bromide (CTAB) and triton (T) [15, 37, 44, 45]. The examples of the effects of different capping agents on the synthesized magnesium borates were presented in **Figure 7**. Kumari et al. [15] reported the characteristic effects of surfactant addition to the liquid-state synthesis of magnesium borate particles and indicated the notable changes in morphology to obtain nano-whiskers [15]. In the synthesis of inorganic ceramic compounds, PVP is utilized to decrease particle size and/or to sustain homogeneous morphology. However, hierarchic porous structures were obtained in the PVP-based synthesis of magnesium borates [44].

3.2 Solid-state (thermal) synthesis

Solid-state synthesis of magnesium borates fundamentally involves the mixture of the powders of raw materials without any liquid component and the reaction of solid powders occurs with the impulsive effect of temperature increase in high-temperature furnaces. The common magnesium sources preferred in this synthesis method are

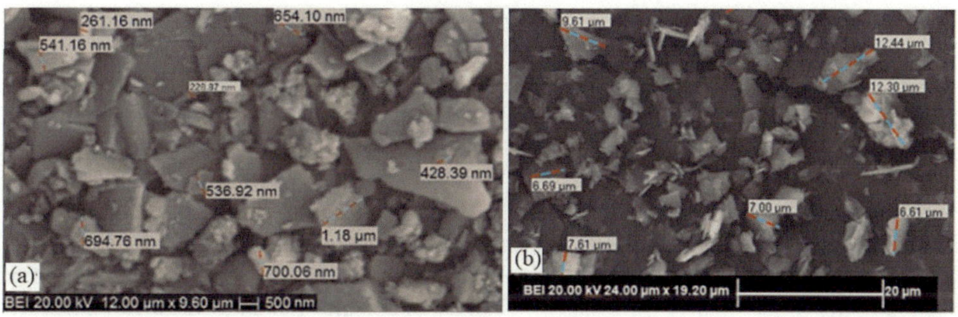

Figure 6.
The comparison of the Admontite morphologies synthesized (a) by ultrasonic liquid-state method [42], and (b) by traditional liquid-state method [6].

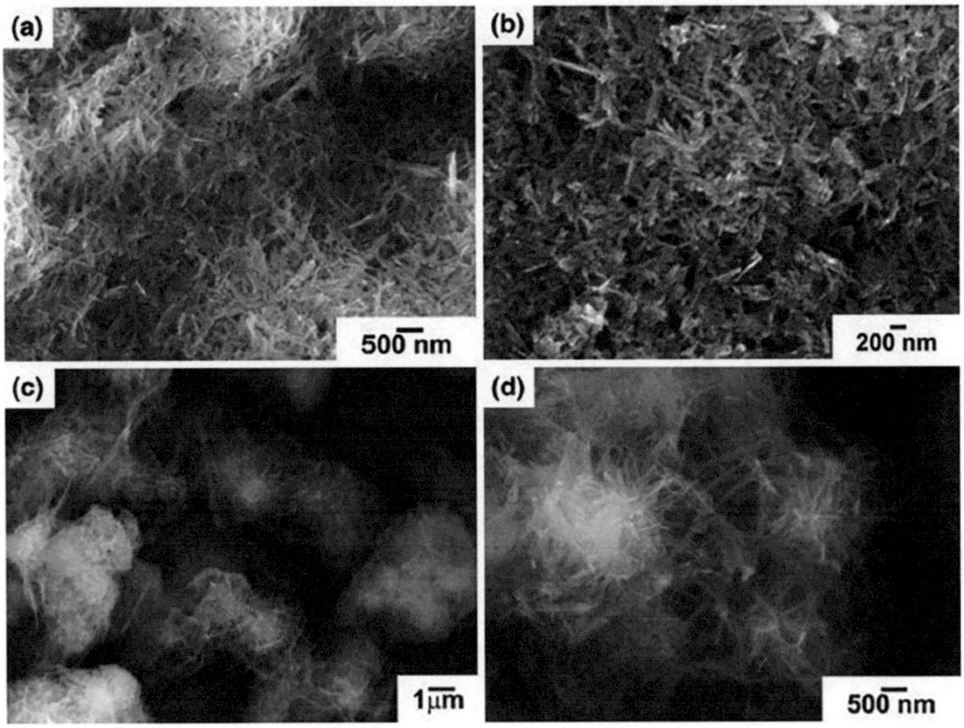

Figure 7.
The effects of different capping agents on the synthesized magnesium borates (a) CTAB, (b) SDS, (c) T at low magnification, and (d) T at high magnification [15].

MgO and magnesium hydroxide ($Mg(OH)_2$). Mostly, the raw materials are reacted in air atmosphere. The prepared sample is commonly in micron-scale at heterogeneous morphology. $Mg_2B_2O_7$, MgB_4O_7, and $Mg_3B_4O_6$ are notable combinations of dehydrated magnesium borates [45, 46].

In the phase diagram of Liu et al. [40] for the solid-state synthesis of magnesium borates, the reaction commonly occurs at higher reaction temperatures than 800°C [40]. In this synthesis procedure, the main drawback of solid-state is the requirement of grinding and sieving processes after the solid-state reaction. Chen et al. [47] used capping agent addition of $Ni(NO_3)_2$ to eliminate these extra processes of the solid-state synthesis procedure [47].

3.2.1 Advantages of microwaves in solid-state conditions

Microwave energy can be defined as non-ionizing electromagnetic radiation with frequencies between 300 MHz and 300 GHz [48]. Similar to the effects of sonochemistry in liquid-state conditions, microwaves can be beneficial to solid-state synthesis procedures to increase the interaction between the powders of raw materials. Unlike traditional calcination techniques, the heating direction is from the inside to the outside of the heated sample in microwave conditions. This situation could assist both to increase reaction yield in solid-state conditions and to decrease the by-product formation. The temperature increase is supplied with the microwave effects. The reaction costs can be reduced effectively with the optimization of reaction conditions. In the reaction procedure, microwave power level and microwave treatment time are notable operating parameters. However, the relationship between the microwave parameters and supplied temperature increase has not been comprehensively studied. In that case, more detailed experimental setups should be designed for the determination of the reaction mechanisms.

Very few studies indicated the possible use of microwave energy in magnesium borate synthesis. Kipcak et al. [49, 50], proved the beneficial effects of microwave to decrease the reaction time in magnesium borate synthesis in comparison with traditional calcination techniques [49, 50].

3.3 Hybrid synthesis

Hybrid synthesis procedure can be defined as the combination of liquid and solid-state conditions. The experimental procedure generally begins in hydrothermal conditions and continues with a solid-state step. Pechini, combustion, and sol-gel synthesis techniques can be assumed as examples of hybrid synthesis [51]. The steps of the hybrid procedure could be more complex than the traditional techniques; however, high purity and homogeneous morphology can be obtained. To strengthen the properties of magnesium borates, hybrid methods should be supported with novel technologies.

Gonzalez et al. [27], synthesized the Tm and Ag-doped MgB_4O_7 with the reaction of $Mg(NO_3)_2$ and H_3BO_3 in urea medium, at the calcination temperature range of 750–950°C by using the combustion method [27]. Zhang et al. [37]; preferred the capping agent of N, N, – dimethylformamide nitrate to fabricate the hierarchic porous particles of magnesium borates. The liquid-state reaction occurred at 150°C for 12 hours whereas the solid-state reaction continued at 600°C for 12 hours [37]. Chen et al. [44], fabricated the mesoporous structure of $Mg_2B_2O_7$ in microsphere morphology by adding the SDS to the reaction medium of $Mg(NO_3)_2$ and borax. The two-step process began with the 80°C for 2 hours and continued with 500°C for 4 hours [44]. Wang et al. [52], prepared the fibers of magnesium borates with a two-step reaction. The mixture of $MgCl_2$ and borax is reacted at 80°C for 12 hours and then sintered at 800°C for 6 hours [52]. In the hybrid synthesis of Zhu et al., the high purity of $Mg_2B_2O_5$ was synthesized with the reaction of $MgCl_2$, H_3BO_3, and NaOH [53].

The comparison of the obtained morphological features of hybrid and traditional procedures can be seen in **Figure 8**. The uniform particle formation was obtained in

Figure 8.
The comparison of the obtained particle morphologies (a) hybrid [44], and (b) traditional [6] synthesis.

Method		Advantages	Disadvantages
Liquid-state	Ultrasonic-Assisted	• Decreasing the reaction time • Uniform morphology	• Increasing reaction temperature with increasing ultrasonic-treatment
	Traditional	• Increasing crystallinity	• Longer reaction times • Non-uniform morphology
Solid-state	Microwave	• Decreasing the reaction time • Uniform morphology	• Complex experimental setup requires
	Traditional	• Increasing crystallinity	• Higher reaction temperatures and times • Decreasing reaction yield • Non-uniform morphology
Hybrid		• Modify the morphology • Decreasing the reaction time and temperature	• Decreasing crystallinity

Table 3.
Comparison of traditional and advanced synthesis techniques.

the hybrid synthesis of Chen et al. [44] whereas the heterogeneous morphology can be seen in the traditional hydrothermal synthesis of Derun et al. [6].

4. Conclusion

Magnesium borates are beneficial for many industrial-scale applications. The ease of their synthesis increases the interest in the studies in this field. The properties of synthesized samples are related with their characteristics. This situation requires the modification of the traditional synthesis method with novel technologies. In this chapter, the relationship between the characteristics, properties, and novel technologies was interpreted. The comparative table of traditional and advanced synthesis methods can be seen in **Table 3**. The common points of the advanced syntheses techniques are the increase in contact interfaces between the molecules of starting materials. Decreasing the reaction time and temperature can be obtained at higher reaction yields with the help of effective contact of molecules. The other important advantage is the modification of particle surfaces.

The improved characteristics showed their effects in the applications. The common applications of magnesium borates are limited by mechanical and radiation permeation. However, the use of magnesium borates with the contribution of their redesigned morphologies was begun in adsorption, ion-battery agent, and hydrogen release agent in chemotherapy.

It is expected that the significance of magnesium borates in industrial applications would be expanded with the increase of advanced technologies in magnesium borate synthesis. In this case, the development of modified synthesis techniques with a novel experimental setup is suggested.

Nomenclature

CTAB	Cetyltrimethylammonium bromide
DTG	Differential thermogravimetric analysis
PVC	Polyvinyl chloride

q_M	Maximum adsorbent capacity
PVP	Polyvinyl pyrrolidone
S_{BET}	BET surface area
SDS	Sodium dodecyl sulphate
T	Triton
TG	Thermogravimetric analysis

Author details

Fatma Tugce Senberber Dumanli
Nisantasi University, Istanbul, Turkey

*Address all correspondence to: fatma.senberber@nisantasi.edu.tr

References

[1] Zhou B, Faucher A, Laskowski R, Terskikh VV, Kroeker S, Sun W, et al. Ultrahigh-Field 25Mg NMR and DFT study of magnesium borate minerals. ACS Earth and Space Chemistry. 2017; **1**(6):299-309. DOI: 10.1021/acsearthspacechem.7b00049

[2] Staiger MP, Pietak AM, Huadmai J, Dias G. Magnesium and its alloys as orthopedic biomaterials: A review. Biomaterials. 2006;**27**:1728-1734. DOI: 10.1016/j.biomaterials.2005.10.003

[3] Seeger M, Otto W, Flick W, Bickelhaupt F, Akkerman OS. Magnesium compounds. Ullmann's Encyclopedia of Industrial Chemistry. 2011;**22**:41-78. DOI: 10.1002/14356007.a15_595.pub2

[4] Mindat.org [Internet]. Available from: https://www.mindat.org/chemsearch.php [Accessed: February 09, 2022]

[5] Anishia SR, Jose MT, Annalakshimi O, Ramasamy V. Thermoluminescence properties of rare earth doped lithium magnesium borate phosphors. Journal of Luminescence. 2011;**213**:2492-2498. DOI: 10.1016/j.jlumin.2011.06.019

[6] Moroydor-Derun E, Kipcak AS, Senberber FT, Sari-Yilmaz M. Characterization and thermal dehydration kinetics of admontite mineral hydrothermally synthesized from magnesium oxide and boric acid precursor. Research on Chemical Intermediates. 2015;**41**(853):866. DOI: 10.1007/s11164-013-1237-6

[7] Moroydor-Derun E, Senberber FT. Characterization and thermal dehydration kinetics of highly crystalline mcallisterite, synthesized at low temperatures. The Scientific World Journal. 2014;**2014**:1-10. DOI: 10.1155/2014/985185

[8] Yongzhong J, Shiyang G, Shuping X, Jun L. FT-IR spectroscopy of supersaturated aqueous solutions of magnesium borate. Spectrochimica Acta A. 2000;**56**(7):1291-1297. DOI: 10.1016/s1386-1425(99)00227-9

[9] Li S, Xu D, Shen H, Zhou J, Fan Y. Synthesis and Raman properties of magnesium borate micro/nanorods. Materials Research Bulletin. 2012;**47**(11):3650-3653. DOI: 10.1016/j.materresbull.2012.06.046

[10] Ahmad P, Khan MI, Akhtar MH, Muhammed G, Iqbal J, Rahim A, et al. Single-step synthesis of magnesium-iron borates composite; An efficient electrocatalyst for dopamine detection. Microchemical Journal. 2021;**160**:1-7. DOI: 10.1016/j.microc.2020.105679

[11] Bahl S, Pandey A, Lochab SP, Aleynikov VE, Molokanov AG, Kumar P. Synthesis and thermoluminescence characteristics of gamma and proton irradiated nanocrystalline MgB_4O_7: Dy, Na. Journal of Luminescence. 2013;**134**:691-698. DOI: 10.1016/j.jlumin.2012.07.008

[12] Du A, Zhang Z, Qu H, Cui Z, Qiao L, Wang L, et al. An efficient organic magnesium borate based electrolyte with non-nucleophilic characteristic for magnesium sulfur battery. Energy & Environmental Science. 2017;**10**:2616-2625. DOI: 10.1039/C7EE02304A

[13] Intemann J, Lutz M, Harder S. Multinuclear magnesium hydride clusters: Selective reduction and catalytic hydroboration of pyridines. Organometallics. 2014;**33**:5722-5729. DOI: 10.1021/om500469h

[14] Hu ZR, Lai R, Wang LG, Chen ZL, Chen GX, Dong JX. Preparation and tribological properties of nanometer magnesium borate as lubricating oil additive. Wear. 2002;**252**(5–6):370-374. DOI: 10.1016/S0043-1648(01)00862-6

[15] Kumari L, Li WZ, Kulkarni S, Wu KH, Chen W, Wang C, et al. Effect of surfactants on the structure and morphology of magnesium borate hydroxide nanowhiskers synthesized by hydrothermal route. Nanoscale Research Letters. 2010;**5**:149-157. DOI: 10.1007/s11671-009-9457-9

[16] Zeng Y, Yang H, Fu W, Qiao L, Chang L, Chen J, et al. Synthesis of magnesium borate ($Mg_2B_2O_5$) nanowires, growth mechanism and their lubricating properties. Materials Research Bulletin. 2008;**43**:2239-2247. DOI: 10.1016/j.materresbull.2007.08.022

[17] He C, Shui A, Ma J, Qian J, Cai M, Tian W, et al. In situ growth magnesium borate whiskers and synthesis of porous ceramics for sound-absorbing. Ceramics International. 2020;**46**:29339-29343. DOI: 10.1016/j.ceramint.2020.08.062

[18] Huang X, Xiao X, Wang X, Yao Z, He J, Fan X, et al. In-situ formation of ultrafine $MgNi_3B_2$ and TiB_2 nanoparticles: Heterogeneous nucleating and grain coarsening retardant agents for magnesium borate in Li-Mg-B-H reactive hydride composite. International Journal of Hydrogen. 2019;**44**:27529-27541. DOI: 10.1016/j.ijhydene.2019.08.222

[19] Bai M, Hu L, Liang Y, Hong B, Lai Y. Enhanced electrochemical properties of lithium-rich cathode materials by magnesium borate surface coating. ChemistrySelect. 2021;**6**:2446-2455. DOI: 10.1002/slct.202004829

[20] Jankowski P, Li Z, Zhao- Karger Z, Diement T, Fichtner M, Vegge T, et al. Development of magnesium borate electrolytes: Explaining the success of $Mg[B(hfip)_4]_2$ salt. Energy Storage Materials. 2022;**45**:1133-1143. DOI: 10.1016/j.ensm.2021.11.012

[21] Fan M, Wen Y, Ye D, Jin Z, Zho P, Chen D, et al. Acid-responsive H2-releasing 2D MgB2 nanosheet for therapeutic synergy and side effect attenuation of gastric cancer chemotherapy. Advanced Healthcare Materials. 2019;**8**:1-9. DOI: 10.1002/adhm.201900157

[22] Ma YQ, Liu ZH. Excellent adsorption performance for Congo red on hierarchical porous magnesium borate microsphere prepared by a template-free hydrothermal method. Journal of the Taiwan Institute of Chemical Engineers. 2018;**86**:92-100. DOI: 10.1016/j.jtice.2018.02.015

[23] Souza LF, Caldas LVE, Junot DO, Silva AMB, Souza DN. Thermal and structural properties of magnesium tetraborate produced by solid state synthesis and precipitation for use in thermoluminescent dosimetry. Radiation Physics and Chemistry. 2019;**164**:1-5. DOI: 10.1016/j.radpyschem.2019.108382

[24] Prokic M. Individual monitoring based on magnesium borate. Radiation Protection Dosimetry. 2007;**125**(1–4): 247-250. DOI: 10.1093/rpd/ncl116

[25] Kumar J, Kumar S, Shekhar C, Brajpuriya R, Vij A. Effect of Eu doping on the thermoulminescence of UV and gamma irradiated $Mg_2B_2O_5$ nanophosphors. Luminescence. 2022: 1-19. DOI: 10.1002/bio.4197

[26] Souza LF, Vidal RM, Souza SO, Souza DN. Thermoluminescent dosimetric comparison for two different MgB_4O_7: Dy production routes. Radiation Physics and Chemistry. 2014;

104:100-103. DOI: 10.1016/j.radphyschem.2014.04.036

[27] Gonzalez PR, Avila O, Mendoza A, Escobar AL. Effect of sintering temperature on sensitivity of MgB_4O_7: Tm, Ag obtained by the solution combustion method. Applied Radiation and Isotopes. 2021;**167**:1-4. DOI: 10.1016/j.apradiso.2020.109459

[28] Pagliaro F, Lotti P, Battiston T, Comboni D, Gatta GD, Camara F, et al. Thermal and compressional behavior of the natural borate kurnakovite, $MgB_3O_3(OH)_5 \cdot 5H_2O$. Construction and Building Materials. 2021;**266**:1-13. DOI: 10.1016/j.conbuildmat.2020.121094

[29] Oduko JM, Harris SJ, Stewart JC. Magnesium borate: Some advantages and disadvantages for practical dosimetry. Radiation Protection Dosimetry. 1984;**8**(4):257-260. DOI: 10.1093/oxfordjournals.rpd.a083062

[30] Prokic M, Christeen P. Graphite mixed magnesium borate TL dosemeters for beta ray dosimetry. Radiation Protection Dosimetry. 1983;**6**(1–4): 133-136. DOI: 10.1093/oxfordjournals.rpd.a082886

[31] Pellicioni M, Prokic M, Esposito A, Nuccetelli CP. Energy response of graphite-mixed magnesium borate TLDs to low energy X-rays. Applied Radiation and Isotopes. 1991;**42**(11):1037-1038

[32] Gao X, Chen Y, Chen P, Xu Z, Zhao L, Hu D. Supercritical CO_2 foaming and shrinkage resistance of thermoplastic polyurethane/modified magnesium borate whisker composite. Journal of CO_2 Utilization. 2022;**57**:1-13. DOI: 10.1016/j.jcou.2022.101887

[33] Zhang L, Zhang M, Chen Z. Study on mechanical and thermal expansion properties of oxide coated magnesium borate whiskers reinforced aluminum-based composite. IOP Conference Series: Earth and Environmental Sciences. 2021; **634**:1-7. DOI: 10.1088/1755-1315/634/1/012097

[34] Baghebaradi MH, Naghizadeh R, Rezaie H, Vostakola MF. Synthesis of dehydrated magnesium borate powders and the effect on the properties of MgO-C refractories. Journal of Ceramic Processing Research. 2018;**19**(3): 218-223. DOI: 10.36410/jcpr.2018.19.3.218

[35] Loiland JA, Zhao Z, Patel A, Hazin P. Boron-containing catalysts for the oxidative dehydrogenation of ethane/propane mixtures. Industrial and Engineering Chemistry Research. 2019; **58**:2170-2180. DOI: 10.1021/acs.iecr.8b04906

[36] Songfeng E, Wu L, Li C, Zhu Z, Long X, Geng R, et al. Growth of boron nitride nanotubes from magnesium diboride catalysts. Nanoscale. 2018;**10**: 13895-13901. DOI: 10.1039/C8NR03167C

[37] Zhang Z, Zhu W, Wang R, Zhang L, Zhu L, Zhang Q. Ionothermal confined self-organization for hierarchical porous magnesium borate superstructures as high efficient adsorbents for dye removal. Journal of Materials Chemistry A. 2014;**2**:19167-19179. DOI: 10.1039/C4TA03580A

[38] Guo RF, Ma YQ, Liu ZH. Three hierarchical porous magnesium borate microspheres: A serial preparation strategy, growth mechanism and excellent adsorption behavior for Congo red. RSC Advances. 2019;**9**: 20009-20018. DOI: 10.1039/C9RA03654G

[39] Zhang W, Wu H, Zhou N, Cai X, Zhang Y, Hu H, et al. Enhanced thermal

stability and flame retardancy of poly (vinyl chloride) based composites by magnesium borate hydrate-mechanically activated lignin. Journal of Inorganic and Organometallic Polymers and Materials. 2021;**31**:3842-3856. DOI: 10.1007/s10904-021-02019-9

[40] Liu Z, Yu J, Wang X, Zhang X, Wang J, Jia D, et al. Molten-salt assisted synthesis and characterization of $Mg_2B_2O_5$ and $Al_{18}B_4O_{33}$ whiskers. Journal of Asian Ceramic Societies. 2021; **9**(3):1298-1309. DOI: 10.1080/21870764.2021.1972591

[41] Chatel G. How sonochemistry contributes to green chemistry. Ultrasonics Sonochemistry. 2018;**40**: 117-122. DOI: 10.1016/j.ultsonch.2017.03.029

[42] Yildirim M, Kipcak AS, Moroydor DE. Sonochemical-assisted magnesium borate synthesis from different boron sources. Polish Journal of Chemical Technology. 2017;**19**(1):81-88. DOI: 10.1515/pjct-2017-0012

[43] Kipcak AS, Moroydor- Derun E, Piskin S. Synthesis and characterization of magnesium borate minerals of admontite and mcallisterite obtained via ultrasonic mixing of magnesium oxide and various sources of boron: A novel method. Turkish Journal of Chemistry. 2014;**38**:792-805. DOI: 10.3906/kim-1307-61

[44] Chen AM, Gu P, Ni ZM. 3D flower-like magnesium borate microspheres assembled by nanosheets synthesized via PVP-assisted method. Materials Letters. 2012;**68**:187-189. DOI: 10.1016/j.matlet.2011.10.041

[45] Storti E, Roso M, Modesti M, Aneziris CG, Colombo P. Preparation and morphology of magnesium borate fibers via electrospinning. Journal of the European Ceramic Society. 2016;**36**: 2593-2599. DOI: 10.1016/j.jeurceramsoc.2016.02.049

[46] Kipcak AS, Yilmaz-Baysoy D, Moroydor-Derun E, Piskin S. Characterization and neutron shielding behavior of dehydrated magnesium borate minerals synthesized via solid-state method. Advances in Materials Science and Engineering. 2013;**2013**:1-9. DOI: 10.1155/2013/747383

[47] Chen S, Zhang D, Sun G. In situ synthesis of porous ceramics with a frame work structure of magnesium borate whiskers. Materials Letters. 2014; **121**:206-208. DOI: 10.1016/j.matlet.2014.01.064

[48] Clark DE, Folz DC, West JK. Processing materials with microwave energy. Materials Science and Engineering A. 2000;**287**:153-158. DOI: 10.1016/S0921-5093(00)00768-1

[49] Kipcak AS, Derun EM, Piskin S. Magnesium borate synthesis by microwave energy: A new method. Journal of Chemistry. 2013;**2013**:1-7. DOI: 10.1155/2013/329238

[50] Kipcak AS, Gurses P, Kunt K, Derun EM, Piskin S. Magnesium borate synthesis by microwave method using $MgCl_2.6H_2O$ and H_3BO_3. International Journal of Materials and Metallurgical Engineering. 2013;7(5):290-295

[51] Lima HRBR, Nascimento DS, Sussuchi EM, Errico F, Souza SO. Synthesis of MgB_4O_7 and $Li_2B_4O_7$ crystals by proteic sol–gel and Pechini methods. Journal of Sol-Gel Science and Technology. 2017;**81**:797-805. DOI: 10.1007/s10971-016-4249-z

[52] Wang X, Peng L, Hua H, Liu Y, Zhang P, Zhao J. Magnesium borate fiber coating separators with high lithium-in